Construction
Administration for
Architects

About the Authors

GREG WINKLER, AIA, LEED AP, is a principal in Green Lizard Management, a project management firm that provides owner representation and project management services to corporate, developer, and institutional clients. An architect with over 25 years of experience in affordable housing, office buildings, industrial/commercial, and retail construction, he was a contributor to the Southern Solar Energy Center publication *The Passive Solar Design Handbook*.

GARY C. CHIUMENTO, Esq., is the founder of Chiumento McNally LLC, a firm specializing in construction-related law. The firm's clients include architects, engineers, surveyors, construction management companies, and various construction professionals. He has published numerous articles on topics of interest to the design and business community and is a frequent speaker on professional liability issues.

Construction Administration for Architects

Greg Winkler, AIA, LEED AP

Gary C. Chiumento, Esq.

New York   Chicago   San Francisco   Lisbon   London   Madrid
Mexico City   Milan   New Delhi   San Juan   Seoul
Singapore   Sydney   Toronto

The McGraw-Hill Companies

Library of Congress Cataloging-in-Publication Data

Winkler, Greg.
   Construction administration for architects/Greg Winkler, Gary C. Chiumento.
      p.   cm.
   Includes index.
   ISBN 978-0-07-162231-8 (alk. paper)
   1. Building—Superintendence. 2. Construction industry—Management.
   3. Project management.   I. Chiumento, Gary C.   II. Title.
   TH438.W565   2009
   690.068—dc22   2009018322

ISBN 978-0-07-162231-8
MHID 0-07-162231-4

Sponsoring Editor   **Copy Editor**
Joy Bramble Oehlkers   Jacquie Wallace, Lone Wolf Enterprises, Ltd.

Production Supervisor   **Proofreader**
Richard C. Ruzycka   Leona Woodson, Lone Wolf Enterprises, Ltd.

Editing Supervisor   **Art Director, Cover**
Stephen M. Smith   Jeff Weeks

Acquisitions Coordinator   **Composition**
Michael Mulcahy   Lone Wolf Enterprises, Ltd.

Project Manager
Jacquie Wallace, Lone Wolf Enterprises, Ltd.

Printed and bound by RRDonnelley.

McGraw-Hill books are available at special quantity discounts to use as premiums and sales promotions, or for use in corporate training programs. To contact a special sales representative, please visit the Contact Us page at www.mhprofessional.com.

This book is printed on acid-free paper.

To my hard-working and dedicated brothers and sisters in the architectural profession.

Greg Winkler, AIA, LEED AP

To my father, who taught me the value of education and hard work. To my mother, who taught me the value of a dollar. To my wife, whose love and support inspire me daily. To my children, who have come to be my friends. To Frank Vitetta, Alan Hoffman, Michael Minton, Richard Holland, the Becicas, Mitch Franken, Rick Piccolli, Peter Coote, Ben & Beth Kitchen, Jack Bauerle, and Keith Palmer, who taught me the business and who always held up to me a shining example of professionalism.

Gary C. Chiumento, Esq.

Contents

CHAPTER 4
Bidding and Negotiation 91

CHAPTER 7
Under Construction—Problems and Disputes 177

CHAPTER 10
Risk Management 245

Preface

A rchitecture is a noble profession, but can be a difficult vocation. Despite the pure joy of design and the satisfaction of seeing those ideas come to fruition, the work is demanding. Owners often expect to receive flawless construction documents quickly, and the contractors and their subs scrutinize the drawings and specifications to find gaps and change order opportunities. If the project takes the path where RFIs, change orders, and accusations are flying, the experience is a strain on all parties. When the project climate deteriorates even further into arbitrations, depositions, and lawsuits, it can be draining and discouraging as well. It is no wonder that some architects view their creations with a mix of pride and regret.

For most of the project, the architect and owner work closely to design and document the building. Whether a major corporate headquarters or an undistinguished warehouse, the building represents an important step forward for the owner, and he is relying on the architect's skills to deliver the project within his budget and schedule. During this period, the architect is the leader of the process, and the acknowledged expert on construction matters in the eyes of the owner.

When a contractor is hired, the relationship fundamentally changes. The contractor becomes the construction master and the leader and initiator of everything affecting the project, and the architect moves into a supporting and secondary role.

Most of the difficult times of architectural practice come during construction administration, when the architect is responding to the needs of the contractor and the strength of the documents is tested daily. It is a time when events are being driven by others and the architect may see himself largely in a reactive mode, responding to the winds of issues blowing in from the work site.

This book argues that architects need not feel themselves victims of fate, and can minimize their pain and suffering during construction administration by better controlling events leading up it, and responding strategically when unforeseen events do occur. It examines construction administration from the architect's perspective: where does risk to the architect originate, and how can we minimize those sources of risk? Risk management is an often overused term, but it has been proven across a number of professions that liability can be controlled through knowledge and planning. Architecture is no different. This book is intended to give the architect the tools to manage his risk during construction administration.

A few notes regarding the tone of this book:

- Without clients there would be no architecture. Owners conceive the initial idea, and architects often personally bond with the owner during the long weeks or months of turning the idea into a tangible, buildable set of documents. We do not diminish the value of such a relationship, but we do argue that at the heart of any project is a business agreement. Even the most agreeable of owners can turn quickly if he comes to believe that the architect has not served his interests well or placed him in financial jeopardy. Where this book may seem to promote a more formal relationship with the owner, that is not our intent. We do, however, argue for a more business-like relationship with the owner.

- Most contractors are honest, hard-working, and more interested in preserving their reputation as quality builders than in hammering the owner for unwarranted change orders. Unfortunately, the less savory variety of contractors are out there, and this book, out of necessity, focuses on the worst attributes of those few.

- We use the masculine gender throughout this book in referring to both architects and contractors. This writing convention is used to avoid the stilted use of *he/she* and *him/her* repetitively. The use of this standard is not intended to minimize the importance of women in any profession, particularly their prominent role in the architectural profession. It is only intended to make the reading a little easier.

Acknowledgments

F or his insightful review of the proposal and manuscript, we thank Bob Ignarri, AIA, of Ignarri-Lummis Architects in Cherry Hill, New Jersey.

For his sharp-eyed editorial and stylistic review, we thank Rob Nigro, ace editor and leader of the Whatever's Write Writers' Group in Haddon Heights, New Jersey.

We thank Joy Bramble Oehlkers, Senior Editor, McGraw-Hill Professional, for her proposal critiques and support.

We also appreciate the editorial and graphic talents of the folks at Lone Wolf Enterprises, including Roger and Leona Woodson, Virginia Howe, and Jacquie Wallace.

1

The Construction Process

A wise old architect once said, "The best buildings are the ones that never get built." He was wise because he realized an essential truth about architectural practice: the greatest liability threat to an architect occurs during the period when his design is under construction. It is perhaps ironic that this is also the period when the architect has collected approximately three-fourths of his fee for the project and his attentions are focused elsewhere, on marketing or preparing the construction documents for new projects. Some architects view construction administration as a necessary evil, a period of trial they must endure to close out a project and see their creation come to life. They may view the contractor with a range of emotions, from a partner in construction to an evil, money-grubbing entity intent on destroying their reputation and their relationship with their client.

The reality, as with most endeavors, is somewhere in-between. Contractors are largely interested in building a quality structure quickly and profitably and moving on to the next project. Few of them are intent on embarrassing or destroying an architect's reputation, for one simple reason—it is not good business. They are also not in business to lose money, however. They will look to recoup the cost of work they do not believe is a part of their contract with the owner. Their main line of attack will go directly through the architect to his construction documents. That is why it is essential that architects view construction administration as the most perilous phase of their work, and take steps to minimize the risks that accompany the act of building what they have drawn.

CONSTRUCTION ADMINISTRATION IS CONTRACT ADMINISTRATION

Construction administration, at its heart, is contract administration. The architect has an agreement with the owner to design a building that meets the owner's needs and budget. The contractor has a contract with the owner to build that building. Between those two simple concepts, of course, exists all manner of complications and difficulties. When construction begins, the architect becomes the owner's representative in assisting the contractor in understanding the construction documents. He also becomes the owner's agent in helping to ensure that the contractor is building to the scope and quality called for in the documents. The owner may or may not view the architect as "master builder" in the great tradition of cinema and the larger-than-life architects of history. But he should certainly view the architect as master of his domain—the construction documents. When the construction documents' completeness and clarity is questioned, the architect must become the interpreter and defender of the documents on behalf of the owner and himself. Construction administration often requires the architect to be part negotiator, part attorney, and part policeman. Mastering these seemingly contradictory roles requires either hard-earned experience or a guide. This book is that guide, wrought from some hard-earned experience.

The purpose of this book is to help the architect negotiate the treacherous shoals of construction administration. We hope it will save the architect—either the rising professional or the experienced hand—much of the pain of learning in the field through trial and error. The first step begins with understanding the basic nature of the contract and the process of creating a finished building.

PRIVATE- VERSUS PUBLIC-SECTOR CONSTRUCTION

The two primary sources of construction in this country are the private and the public sectors. The private sector of the economy is responsible for the overwhelming majority of construction projects in residential, commercial, industrial, and heavy construction. Federal, state, and local governments underwrite a significant amount of construction in the public interest and for the public use; these projects are sponsored and paid for with taxpayer money (or funds borrowed based on the guarantee of a taxing authority). The state and federal statutes and rules regulating the procurement of construction and construction-related services mark the major distinction in private versus public construction. Professional services such as architecture, engineering, and construction/project management are rarely selected by strict price competition or bidding. Typically the design professional responds comprehensively to a *Request for Proposal* (RFP), a detailed identification of the project program and its financial parameters. The selection process is often subjective, befitting the artistic and aesthetic considerations influencing the selection of the design professional by the government entity.

This does not mean, however, that the design professional is immune from government regulation, or the ubiquitous red tape that accompanies it. Because of the myriad of regulations created by the various federal agencies, the 50 states, and the thousands of public bodies that have some responsibility for building capital projects, it is beyond the scope of this book to discuss in any detail the responsibilities of the architect and engineer in the public sector. This is especially true where contracts for professional design and construction administration services can be offered and awarded on a non-competitive basis. This can trigger regulations regarding *pay-to-play* prohibitions, which are designed to minimize influence exerted by vendors (such as design professionals) who contribute to state or local-level office holders who are able to exert considerable influence in the award of architectural commissions in large governmental capital projects.

Public-sector construction can also be a fertile ground for public policy legislation often unknown in the private-sector. For example, *no damage for delay*, anti-indemnity provisions, and other limitations as to liability are far more prevalent in public-sector work than in the private sector. Additionally, competitive bidding (which often encourages cut-throat practices and change order mania), and the political considerations of construction often place the design professional at greater risk for claims than is commonly seen in private-sector projects. The architect who competes in the public arena for commissions should familiarize himself with the rules of the game, including: Federal Acquisition Regulations, state contracting statutes, and the specific requirements of the jurisdiction where the project is constructed. Relevant differences between public and private projects will be noted throughout this book.

THE AGREEMENTS

The agreement is the glue that binds two parties to each other in a relationship that provides something of value to each party. The legal foundation, the chemical bond of the glue, for the relationship between all parties in commercial transactions is not the desire to contract, but the need that each has for what the other has to offer. The modern concept of contract results from the recognition that the complexity of relationships demands that an agreement be memorialized—or written down—or potentially lost. Everyone recognizes that the dullest pencil is sharper than the sharpest memory, but our disdain for formality often overcomes recognition of the need for a formal written document that lists what one party will do for the other, and at what price. While it is not the primary purpose of this book to discuss all the details of effective contract negotiation and recording, it is essential to start the discussion of effective contract administration by identifying the importance of the written agreement between the owner and the architect, and the provisions that must be included to ensure enforceability.

The Importance of Written Agreements

Attorneys are fond of telling clients in contractual disputes: "If it isn't written; it didn't happen." This can be paraphrased to apply to written agreements as well. Good contracts state the requirements of each party. More specifically, good contracts define what the architect is, and is not, responsible for providing to an owner. Often the contract takes the form of an exchange of oral promises. The architect will promise to provide certain services in exchange for a promise to pay under certain terms and conditions. This exchange of promises (rather than the actual services or payments themselves) is, in fact, the contract and is under most circumstances enforceable by a court. Certain types of contracts must be in writing as required by the Statute of Fraud of most states, so called because the requirement of writing is present to prevent frauds. Most architectural agreements can, conceivably, occur as oral contracts, but it is almost a truism that the architect's oral agreement isn't worth the paper it isn't written on (and many states now state that only written contracts are enforceable). This is not because the architect is any less reliable or trustworthy than any other party to a contract. It simply results from the twin human failings of differing understandings and the vagaries of memory. Even the most sincere and collegial team of owner and client will understand and remember verbal agreements differently. Even written agreements are often the source of disagreements in court, with each side parsing the language to argue for a favorable meaning of the verbiage. Verbal agreements offer very little protection to the architect. Architects who provide services based solely on oral promises often argue that they prefer not to be bound by the detailed intricacies of a written agreement. The answer to this argument is that the unwritten expectations of clients (and contractors) may be far beyond what an architect can reasonably provide. For every excess obligation created by a written owner/architect agreement, a large number of unreasonable or absurd expectations are removed. Standard forms of agreements available from the American Institute of Architects (AIA) carry with them long histories of case law that define limits and reasonable expectations of professional service. Even simple professional agreements that define the scope of work and the services provided by the architect provide a substantial level of protection over verbal agreements. Providing architectural services, even for small or limited projects, without a written agreement is simply foolish. Since the availability of technology allows any architect to prepare some form of written agreement, there is no excuse not to have a written contract, even if only a rudimentary one.

Types of Agreements

Two principal types of agreements are used between owners and architects: the standard form agreement and homemade agreements.

Form agreements are available from a variety of sources for commercial contracts. Most of these sources provide a "bare bones" approach, and include the listing of the exchange (what A will do and what B will pay) and some basic terms and conditions. The architect should consider as an alternative to this approach the use of standard form agreements sponsored by such organizations as the American Institute of Architects, the American Society of Civil Engineers, and the Engineers Joint Contract Documents Committee (EJCDC). Each of these organizations takes a "family of documents" approach to contracting, with the goal of providing a comprehensive set of contracts that are fair and balanced while addressing in neutral fashion the likely topics the owner and design professional will want in their agreements. Both organizations also produce a line of documents that are supported by civil case law across the nation. The greatest benefit of this approach is the likelihood that most of the elements the architect wants in his contract will be included and not left out simply because the parties forgot to address them. It also serves as a good checklist of important issues that can be adjusted by a rider or directly on the page. The fairly nominal cost for an individual contract is far and away justified by the benefits of comprehensiveness in addressing issues most architects would not even think of, much less remember to include if they were creating an agreement from scratch.

Homemade Contracts

Unlike home cooking, homemade contracts are not usually good for an architect. They suffer from a lack of breadth and detail, much as if mom had left out the dressing and mashed potatoes from a Thanksgiving supper. While it would be extreme to suggest that a bad contract is better than no contract at all, it is often a close call. A good contract should at the very least serve the purpose of defining explicitly what services the architect will provide, and thereby define by omission the services he will not provide and for which he cannot be held responsible. The most common form of homemade agreement is the countersigned proposal. In this type of agreement, the architect responds to the request of the owner for a proposal including what he would do and how much he would charge. See the C.A. Anecdote for an anecdotal example of the hazards of homemade contracts.

Architects and owners share a special relationship during the design and construction of a building. In most ways, it is a symbiotic relationship: the owner pays the architect money for a service. It can be much more than that, however. Even the owner of the most mundane and purely functional building has his own ideas wrapped up in it. Most owners know exactly what they want from their new building. They may not all express it clearly, but they understand that they need an architect to help them achieve those goals. What the architect needs from the owner, more than any other single piece of information, is to clearly understand his expectations.

Every proposal should include a detailed statement of the programs and services the owner expects of the architect from planning to design, including creation of construction documents, responsibility (or not) for permits and other regulatory requirements, procurement of construction contractors, and, most importantly, a detailed description of services expected by the owner once construction begins. The contract should clearly identify how the parties will handle requests for services outside the identified scope and, if appropriate, identification of the services most likely to be considered *extras*. Listed below are some recommended minimum provisions for any agreement between an owner and architect.

RECOMMENDED OWNER-ARCHITECT CONTRACT PROVISIONS

1. State the acts that constitute default of the contract: *Examples include the owner's failures to make timely payments and the architect's failure to deliver a full set of construction documents. This section should also state the parties' rights and remedies when the other defaults.*

2. State the payment terms and conditions: *What is the compensation for each phase of work? How often does the architect bill? How many days does the owner have to pay? What interest rate does the architect charge if the owner pays late?*

3. Stipulated alternate dispute resolution methods: *Are mediation and arbitraion allowed as alternatives means of resolving disputes?*

C.A. Anecdote

Handshake Agreement Costs Friendship

The Problem

Bill read the letter from his old friend with a sinking feeling. He had known Charlie for more than twenty years. They had met in the local Rotary Club, vacationed together a couple of times, and shared a golf course frequently. When Charlie wanted to build a large service bay addition to his car dealership he asked his architect friend Bill to help.

Bill told Charlie he would provide the schematic and design development services on an hourly basis, charging only for his staff time. For the construction document phase, Bill agreed to give Charlie a fixed fee. Charlie was concerned about the overall costs of the architectural services, even though Bill was providing his own time at no cost. Bill assured him he would treat him fairly and design a building he could afford. They shook hands and Bill launched the project the next day. When the bids came in high, Charlie was distraught. "This happens all the time," Bill told him. "Relax. We'll sit down with the low bidder, cut some scope, and get it back within your budget."

Now he was reading a letter from Charlie telling him that his overdue invoice for the bulk of the construction documents would not be paid, and that Charlie was referring the matter to his attorney to reclaim all fees paid on the project. "Despite your assurances to the contrary," read the letter, "the project as designed cannot be constructed within the approved budget. You are therefore in default of our contract."

The Resolution

Bill's attorney handed the letter back to him.

"Charlie didn't write this," he said. "He had an attorney write it for him."

"So much for old friendships," said Bill.

"Where is the contract he refers to in the letter?"

"There isn't one. We just agreed and shook hands. A contract seemed too much given our history. What do we do now?"

"Without a signed agreement, both parties are making their claims on verbal agreements. You have an argument that he owes you the remainder of the fee, but he has an argument that you owe him a building that can be constructed within his budget."

"So how do I get my money?" asked Bill.

"You don't. Each of you can spend a lot on legal fees and you'll prob-ably end up right back where you are now. He'll have a set of construc-tion documents he can't use and you won't be paid for all your work. Here's my recommendation: I'll write him and suggest both parties drop their claims against each other and move on."

"I'm losing a friend and a fee," Bill said. "This is an expensive lesson."

4. Stipulate verbal integration/written modifications only: *This means the written agreement supersedes all previous discus-sions or versions, and the written agreement can only be changed in writing, signed by all parties.*

5. Stipulate that the owner agrees to a Waiver of Subrogation: *Insert a statement that no party or its insurance carrier can sue for insured losses.*

6. State when the statute of limitation begins: *Most states have a statute of limitation (or statute of repose) that limits the amount of time an architect can be sued for professional ser-vices on a project. Every agreement should state that this time period commences at the date of substantial completion.*

7. Limitation of liability: *State a limitation of liability equal to the total fee for the project or, if the owner objects, at least the limits of the architect's professional liability coverage.*

8. Waiver of consequential damages: *Consequential damages include: lost rents, damage to reputation, down or idle time, interest and finance charges, loss of use of goods, additional labor costs, material escalation costs, depreciation, rental costs, additional energy costs, and loss of productivity or effi-ciency. If an architect is determined by a court to have liabil-ity for consequential damages, he can be assessed damages many times the fee he was paid for the project. Standard AIA contracts include a mutual waiver of these damages between owner and architect.*

Even in the case of homemade agreements, an architect or engineer can benefit from a trip to an attorney with experience representing design professionals. The attorney can identify a wide variety of clauses that protect the architect from unfair expectations of the owner. These clauses can come in the form of a set of boilerplate terms and conditions which can easily be incorporated by reference in the proposal submitted by the architect. If the design professional does not know an attorney who is skilled in construction law, he can often refer to his professional liability (PL) insurance carrier or his broker, who can provide information as to the availability of appropriate contracts that conform to the requirements of their PL policy of insurance. Often, an architect's professional liability broker or insurance company will offer to review an owner/architect agreement, particularly in cases where the architect is concerned because the owner is including unusual or onerous clauses that may be beyond the architect's insurance coverage. This type of review can be invaluable in helping the architect to reject clauses that his insurance policy will not cover, or to craft new contractual language that meets the owner halfway while still preserving insurance protection for the architect.

THE TRIANGLE

The architect must be able to recognize and work within various forms of building delivery systems. The most common forms of delivery systems are the *Design-Bid-Build (DBB)* and the *Design-Build (DB)* methods. Each method carries different obligations, goals, and pitfalls for the architect (see Risk Hazard Flag). Understanding, contractually, how a project will be constructed is key to the management of risk during construction.

Design-Bid-Build (DBB)

In the DBB method, the architect typically provides designs in the form of plans and specifications in support of the owner's program, and also provides varying degrees of assistance in procuring a contractor who is solely responsible to erect the edifice. Usually the architect or engineer is careful to steer clear of taking responsibility for building the project, deferring to the acknowledged expert in construction—the contractor. It is not unusual for the architect to be involved in the construction phase by providing administration of the construction agreement, also known as construction administration (CA). While this term will be defined in detail and explained later in this book, suffice it to say that CA is the process of clarifying and explaining the design to assist the contractor in constructing the project in adherence to the construction documents. Owners like the DBB method because they feel that bidding out the construction documents to a number of contractors assures them of receiving the most competitive pricing. Architects recognize that the act of bidding, while common, exposes their construction documents to intense scrutiny by a number of contractors. While the questions generated by such a wide review often enable the architect to clean up discrepancies, tighten unclear areas, and otherwise strengthen the successful bidder's understanding of the contract—it can be stressful, and an indication of underlying problems in the construction documents.

Design-Build (DB)

Design-build (DB) is a delivery system that often removes the barrier between design and construction and makes the architect and the contractor teammates in a joint effort to build the project for the owner. The benefits of the relationship to the owner include a more collegial atmosphere on the project and less finger-pointing in the event of deficiencies, construction delays, personal injury, or property damage. However, there are obvious disadvantages as well, including: a blurring of the design professional's responsibility to identify deficiencies during construction, and the loss of constructability or earnest value-engineering efforts at the commencement of the project. Owners appreciate that design-build contracts often save them time in the documentation and bidding phases of a project, and remove the occasional disputes between architect and contractor over what is reasonably included in the construction documents. Owners do not always understand that they lose the architect as advocate in this type of project. It is often argued that the architect's financial responsibility shifts from the owner to his partner—the contractor.

While design professionals should consider seriously the pros and cons of design-build, this book assumes that the project under consideration (and under construction) is a conventional design-bid-build project, with the architect or engineer hired by the owner to design the project, assist the owner in procuring a single prime (or general) construction contractor, and provide administration of the construction contract. See the Risk Hazard Flags for tips on reducing ths risk associated with various types of project delivery methods.

The Triangular Relationship

It is helpful to think of the relationship among owner, architect, and contractor as a triangle (see Figure 1-1). Note that the owner is at the apex, where he is supported by the design professional (architect or engineer) and the contractor on separate legs of the triangle. These solid line legs represent a direct contractual relationship (hopefully in writing) between owner and architect and between owner and contractor. The relationship between

 Risk Hazard Flags

Project Delivery Risk Reduction
Design-Bid-Build
- Requires tight and thorough bid documents
- Thorough review of documents by bidders
- Architect is watch-dog for owner

Design-Build
- Ensure owner knows architect and contractor are a team
- Architect is no longer independent owner's agent

Fast-Track
- Offers the owner schedule/budget rewards and greater risks
- Requires careful foundation and structural assessment by architect

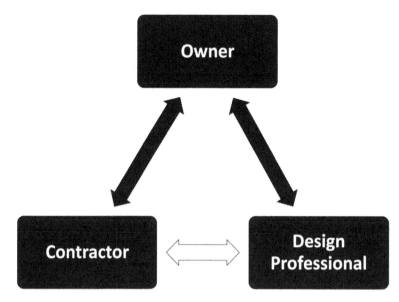

FIGURE 1-1
The owner, architect, and contractor triangle.

architect and contractor is depicted by an open arrow that represents a relationship not likely to be fixed by a contract. Often the relationship is defined in the construction documents by a listing of the roles that each plays in supporting the owner's effort to achieve the desired result. The architect, for instance, has an obligation to review and respond to submittals prepared by the contractor in a timely manner. The contractor, for his part, has an obligation to ensure that his subcontractors and suppliers are preparing the required submittals in conformance with the specifications. Neither the architect nor the contractor has a direct contractual relationship with each other to cooperate in this area, but both are contractually obligated to the owner to do so. This example should remind all participants of their common goal—the support of the owner's program—and their equivalent importance in achieving that goal.

Fast-Track Project

One more variation is worth noting, especially where the design professional will be providing construction administration services—the fast-track project. In this type of project, the owner and contractor believe that economies of scale can be obtained by beginning construction on some parts of the building before total design of the project has been completed. Fast-track construction is often employed in design-build arrangements, where the contractor and architect are in league, but it can be used in standard owner/contractor agreements as well. Often this project is completed in phases, where the owner thinks that time and money can be saved by timing the construction to take advantage of fluctuations in the market for labor, materials, contractor availability, or a favorable regulatory condition. The architect who is involved with overall scheduling and design of the project must include sufficient time to allow for lurches in the schedule due to a miscalculation in any of these areas. The architect may want to consider the retention of a scheduling expert to plan accordingly. There must also be a recognition on the part of all contractors and suppliers of the consequences of unexcused delays to the project. Fast-track construction must also clearly be labeled in the bid drawings or contractor procurement phase of the project. The owner must be thoroughly educated in both the rewards and the risks, and this education must be documented. The owner may not recall the design professional's discussions with him of the risks of fast-tracking when the schedule lags, whether or not this is the fault of the architect. Such timed items as submittal returns, responses to requests for information (RFIs), and payment application reviews, usually critical in any project, take on even greater significance with the increased schedule urgency of fast-track construction.

The Consultant Team

The importance of a solid and reliable team behind each major participant cannot be overstated. The architect must have timely, accurate, and complete design documents to ensure compliance with his contractual obligations to the owner. Because these components are not often under the architect's direct control, he must work to develop strong relationships of trust and confidence in his consultants, and even with consultants who are working directly for the owner. It is wise to develop relationships with several firms to avoid the possibility of delays due to conflicts of interest or lack of capacity to do a particular job. Familiarity with the consultant's personnel (not just with its marketing representative or point person) can help the architect identify and solve problems of delivering the documents not only on time, but also with a high degree of confidence in their constructability. The architect's project manager is responsible for leading the consultant team. This means being the primary professional in a number of key areas: ensuring document delivery schedule obligations are understood, resolving problems, or answering questions from team members to enable them to complete their work; communicating goals, changes, and other issues clearly and quickly; and coordinating the documents among the architect and his consultants.

SUPPORTING PLAYERS

Although not shown in the triangle, supporting players at all points on the triangle play a significant part in achieving the owner's goal. The architect is supported by members of his own staff (drafting support, field representatives, clerical personnel) and professional consultants in numerous design fields, including: mechanical, electrical, plumbing, structural, civil/site, geotechnical, fire protection, kitchen/bath, lighting, acoustical, and other specialty areas. The contractor is supported by subcontractors (in various tiers), including: vendors, estimators, submittal specialists, and insurance brokers. Of course, the owner is not without his team as well, including: project/construction managers, clerks of the works or owner's representatives, lawyers, accountants, marketing and advertising firms, realty or sales associates; and where appropriate, direct design consultants (civil engineering, interior designers, landscape architects). On large projects, the coordination of all the consultants involved on each side can be a true headache. Even communication, when left unmanaged, can result in confusion and consternation when an owner's consultant, for instance, begins making requests of the architectural team that are outside their contractual scope or that affect the schedule of their core responsibilities. To avoid these difficulties, the most successful teams spend time discussing the responsibilities and needs of each member of their team, and establish lines of communication that run through the owner, architectural, and contracting principals.

THE CAST OF CHARACTERS

Knowing the parties to the construction and their respective roles is an important part of understanding the construction process. It can be embarrassing to the young project architect, either in field observation or conducting a job meeting, to have to admit ignorance of the various parties to the construction project or their respective roles. It is especially troublesome where the design professional is partly informed or ignorant about the obligations and responsibilities of his own team. Understanding the roles and responsibilities of each party is also critical for another reason: It enables the architect to know his/her limitations as well. A subcontractor, for example, cannot obligate the contractor to perform work that he does not agree is a part of the contract. Similarly, an architect cannot typically authorize a change to the value of the contract without the owner's approval. Architects who ignore the obligations and limitations of their partners in the construction process expose themselves to the risk of unwittingly absorbing the obligation themselves. Following is a brief review of the role of each cast member in the great comedy (and occasional tragedy) of construction.

The Owner

An architect once remarked that the profession would be thoroughly enjoyable except for the fact that owners are involved. True enough perhaps, except that the owner is the party that sponsors the construction. He often, but not always, owns the ground on which the project is erected. Sometimes the owner is a professional developer who is erecting a turn-key project for a third party who will eventually have a fee-simple or complete ownership interest in the project, such as a single-family home or a commercial office building. Sometimes the project is for something less than fee-simple, such as a long-term tenant in a build-to-suit arrangement. Most notably, developers of community-based living arrangements (such as condominiums, cooperatives, or zero-lot line developments) build multi-family dwellings where an individual owner purchases a unit. In this type of development, the purchaser owns everything inside of the drywall while also owning an undivided percentage of the *common elements*. The least complex method of determining the undivided ownership of common elements is to compare the square footage of the unit purchased to the entire square footage of living space. Common elements usually include:

- Building systems outboard of the drywall (such as structural elements, roof, sheathing/siding, glazing, rough plumbing, mechanical, and electrical).

- Other shared elements of the building envelope (attics, basements, vestibules and entrances, hallways, mechanical and service areas).

- All other elements and property outside the envelope (including roads and walkways, landscaping, accessory buildings, community buildings).

- Recreational features (such as park and playground areas, clubhouses, and pools).

Architects should be particularly careful in providing services for community associations as they form the leading area of housing growth and the largest single group of parties filing design and construction deficiency claims in the nation.

Whether the owner is building for himself or for the ultimate use of another, certain characteristics are the same.

The Program

The owner starts with an idea. Typically the idea is for the use of his property in a more profitable or beneficial way. The use may be fitted to a specific piece of ground due to its location, its topography or its proximity to natural or man-made resources. The owner is probably ill-equipped to turn the idea into a workable program that will one day be a useable facility, however, without help from a designer and builder. To accomplish this goal for the owner, the architect needs to be aware of the information the owner needs to bring to the table and advise the neophyte owner of his responsibilities and how to meet them.

The Location

The owner provides for the physical location of the construction by obtaining an interest in the property either by deed, lease, or license (permission to build) that permits the erection of the building. Problems with this requirement most frequently occur where the owner has failed to obtain a survey of the property (or has obtained a deficient survey) and cannot rely on the boundary lines established. Another related deficiency is where the owner fails either to secure or provide for necessary easements for drainage, utilities, and parking. The consequences of the failure to secure permission are as devastating as they are obvious, and include costs associated with obtaining after-the-fact permission to build on the land of another, costs associated with demolition and rebuilding, and delay associated with the remediation. The architect can protect himself and his client against these issues by insisting on appropriate evidence of permission to build in the form of deeds and surveys attesting to the owner's entitlement to construct on the designated property.

Legal Support of the Project

The owner is ultimately responsible to ensure that legally enforceable documents are in place regarding the essential elements of the project design and construction. Here the architect is especially cautioned to participate in reviewing the contractor's contract with the owner to ensure there are sufficient safeguards for the integrity of the project and his own welfare, both of which have strong benefits to the owner. The specific subject of the necessary terms of the owner-contractor agreement is more fully explored in Chapter 2.

The owner is also responsible for providing the resources necessary to complete the *Work* (that is the legal term for the scope of the project defined in the construction documents), which constitutes the owner's program. Most notably the owner is required to have sufficient financial resources to ensure that the completion of the project will not be jeopardized by a lack of funding, through either cash or financing. It is not at all unusual for the architect to include a provision in his contract requiring the owner to certify his financial ability to continue the project. Contractors may also request proof of the owner's financial ability before they commence work on a project.

Administrative Support

Even the simplest type of construction effort requires administrative support of varying degrees. The regulatory requirements for construction can be annoying and have driven more than one owner to distraction. From obtaining planning and zoning approvals and building permits, to dealing with the bank for financing and obtaining necessary insurances, the owner's plate is filled with numerous responsibilities that belong to neither the architect nor the contractor. And yet the owner is not without help in these and other areas.

The owner can call on any number of professionals and trades, limited only by imagination (and purse), to assist him in the administration of the project. These parties include:

- Engineers to provide civil and site design, and investigate geotechnical and environmental issues
- Lawyers to negotiate and draft contracts and provide representation before regulatory bodies
- Accountants to assist in determining financial status and accountability among the participants
- Insurance brokers to develop the proper amounts and types of insurance and to assist in determining when other project participants are deficient in their insurance requirements
- Project representatives to be the keeper of project records and provide an essential communications link between owner and contractor when the owner is incapable (or unwilling) to do so

- Construction managers, distinguishable from project repre-sentatives by their ability to provide technical assistance in actual construction efforts, inclusive of a variety of tasks, including: Program development, constructability reviews, contractor procurement construction, substantial comple-tion, final completion, and beyond.

Owners have obligations to the professional team and the contrac-tor that they must meet. Conflicts—and risk—can occur where one of the other parties in the triangle of construction assumes these responsibilities through either a sense of obligation or a misguided desire to help the client by *filling the void*. Every architect faces the decision of when to move beyond his contract to maintain goodwill with the client and move the project along. When doing so, he must take the action knowingly and in a manner that best protects his interests.

The Architect/Lead Design Professional

The lead design professional (LDP), while often a registered archi-tect licensed to practice the architectural profession in the state where the project is located, need not be a registered architect depending on the nature of the project and the particular state regulation of such professions. Architects generally design build-ing envelopes for human use or habitation, such as residences and buildings for assembly or commercial use, while engineers provide technical design support for such projects or take the lead in other projects such as industrial and heavy construction design. Some states provide for and permit overlap in these areas, depending on the main purpose of the construction. The design professional should carefully review regulations for practice of the respective profession in the state where the project is located before provid-ing professional services. Interestingly, it may be a violation of that state's criminal law to practice a profession without a specific license. To do so can result in a fine—or even incarceration!

Whether architect or engineer, the lead design professional's role and responsibilities to the project and to the owner are the same. The owner has a legitimate expectation that the lead design pro-fessional will provide services in conformance with standards for design professionals in that community consistent with the contractual scope of work. The contractual scope of work is critical to a determination of the architect's obligations to the project. Usually, it includes three main phases: 1) Design of the project, 2) Procurement of construction contractors, and 3) Construction administration, or assistance to the owner during construction.

It is important to acknowledge the wide degree of variation that exists in determining the architect's design requirements. For instance, the owner may call upon the LDP for total responsibility for programming, designing, detailing, and documenting the design, providing only the vaguest notion of what he wants to do on the selected site. On the other hand, the owner may have a fully formed notion of what he desires for the project, down to the fixtures, colors, and finishes. In these cases, the architect may merely be required to document the project to obtain regulatory approval and building permits. This is most common in prototype residential housing, big-box commercial design, and repetitive franchise retail outlets where the owner is all but leasing the LDP's license, not tapping his creativity or ingenuity. Such projects nevertheless test the skill and the ability of the LDP to deliver the project on time, within the stated budget, and with strict adherence to national, state, and local building codes while faithfully maintaining the owner's vision (if not the architect's) for the project. The drawings must not only provide the project depictions for the builder's use in constructing the project, they serve several other important purposes as well. The documents must be capable of describing the work accurately, so as to serve as the basis for take-offs and other techniques used by owners and contractors to estimate the cost of the work so the contractor can quote a price and the owner can blunt any criticism by the contractor that the construction documents were unreliable and unusable for that purpose. The typical result of such criticism is that the contractor will insist that he is entitled to extra compensation for the missing or poorly depicted work in the construction documents.

Another important purpose of the documents is to secure necessary approvals and permits, most notably the building permit. Many days of delay have been generated (or at least allegedly generated) by the failure of the architect to appropriately document the requirements of state or local building codes in the construction documents. The LDP should include regular quality control reviews of the plans for life, health, and safety requirements of applicable building codes prior to permit submission. Where local code authorities will allow the architect to obtain a preliminary or cursory review for major concerns, he should take advantage of this opportunity.

A related duty of the LDP that is often given short shrift is the coordination of other consultants who will provide technical design drawings, details, and specifications. While the architect may not have had a direct hand in the creation of the designs of other professionals in his service, he will most assuredly be held legally responsible for their deficiencies. Unfortunately, technical inaccuracies and failings are not the only problems. Fragmented or incomplete submissions, or tardy responses to contractor questions and owner modifications, are the architect's responsibility to rectify. Even if the issue is one that requires the architect's consultant to take a lead role in resolving, the architect must take on the task of ensuring the problem is resolved quickly, in a coordinated manner, and in the best interest of the owner. Eternal vigilance in the LDP-consultant relationship is the price for keeping the owner happy and minimizing claims.

The lead design professional's support team usually includes some or all of the following.

Technical Consultants

The nature, type, and number of consultants are determined by the scope of the project and the LDP's in-house capacity. Project consultants can include mechanical, electrical, plumbing, structural, civil, environmental, geotechnical, acoustical, lighting, traffic, fire protection, security, kitchen and bath, interior, and landscape consultants. Strong written agreements between the LDP and his consultants which address the consultants' relationships with the owner (and adopting the agreement at least with regard to the duties and obligations of the LDP to the owner), payment issue, and indemnity in the event of claims or losses are recommended in lieu of a countersigned price and scope proposal.

Schedulers and Estimators

Depending upon the LDP's scope of responsibilities for the project, it may be necessary to use the services of a cost estimator to update regularly the likely cost of the owner's program at the completion of each design phase. Scheduling is usually the chore of the contractor or construction manager, but it sometimes falls to the LDP to identify the probable time for construction, and to ensure the contractor is maintaining the schedule and not wrongfully relying on non-existent *float*. Such float (or excess time in the schedule intended to accommodate schedule overruns) may be a comforting illusion, like an oasis in the desert that quickly dissolves in the face of the real schedule pressures of weather, subcontractor, and inspection delays. The LDP may also be required to act as an arbiter of claims brought by contractors based on delays allegedly resulting from owner delays or errors or omissions in the LDP's work. If the LDP's firm does not regularly provide such services, it should retain someone with technical and local market expertise in these areas—someone knowledgeable of contractor competition, labor, and the costs and quantities of materials in the market area of the project. The consultant should also be familiar with critical path method (CPM) and other scheduling theory, and have the capability to interpret and comment upon a complex construction schedule.

Administrative Support

The LDP can call upon the same battery of support professionals as the owner, but often chooses not to, particularly where additional overhead costs cast a pall on profitability. Resort to an attorney for contract advice or to a professional liability broker for possible claims issues to help nip many annoyances in the bud before they become full-blown problems. The cost for these services may be minimal if dealing with an attorney regularly, or nothing at all in the case of an insurance broker.

The Contractor

The contractor is charged with the responsibility of taking the plans and specifications and, for the agreed upon price, delivering the building in a finished, usable condition, fit for the owner's expressed needs. This must be the end goal of all construction contracts, and anything short of this leaves the owner with little more than a figurative hole in the ground into which he continues to pour money. The selected contractor, whether the result of direct negotiation or open public bidding, must have a substantive opportunity to investigate the site, review thoroughly the contract documents, and match the obligations of the contract and the construction documents to the resources available in the form of his own forces and the subcontractors, vendors, and suppliers to whom he has access. A failure to size up the project for cost of overhead, labor, materials, equipment, and supplies for himself and his likely subcontractors will undoubtedly result either in a bid so high as to guarantee the job goes to someone else or so low as to dictate a loss. Despite the illusion of short-term benefits to the owner from a contractor who suffers a loss on a project, most experienced owners and developers know that a contractor who operates at a loss spells trouble in the form of delays, shoddy workmanship, materials below specification, and ultimately the possibility of *walk-off* contractors and claims for damages. The owner and his architect should always compare contractors' bids and prices to see whether the low bidders have left "too much money on the table," thereby signifying a train wreck of a project that will reveal itself when the contractor is too deeply involved to allow an easy resolution. Indeed, a post-bid debriefing to ensure that there is sufficient coverage of the line items of the project and the contractor's financial ability to sustain the construction is always advisable.

Contractor Obligations

In addition to bidding the actual design specifications, the contractor must be cognizant of the *performance specifications*. Here the architect places in the hands of the contractor complete responsibility, including original design and construction for certain building systems with specified performance criteria as the contractor's guide. Typically, the architect may hand over responsibility for fire protection, accessory use buildings (such as metal storage or utility structures), trussed-roof systems, or even HVAC systems, once firm performance or end-use characteristics have been established. The specifications usually require the contractor to retain a licensed professional engineer to provide the plans for the actual design. The designs are provided as if they are a submittal, and approved in the same fashion by the architect or his consultant. While it is sometimes the case that the architect's consultant is engaged by the contractor to provide this design (even though it is at the contractor's expense) this practice is not recommended. The engineer must consider the complications that can arise if the design of the performance item comes into question and the owner is required to file a claim against the parties responsible for it. Performing as the engineer for both the owner and contractor places the engineer in an untenable position when problems arise between the owner and contractor.

The contractor is responsible for the *means and methods* of construction, as well as its sequencing, scheduling, and all aspects of the actual performance of the work, including the safety both of the work and of the workers. The *front-end* specifications are a critical device to reiterate these requirements, refine them, and close all loopholes so there is no question about the contractor's responsibility.

Pricing the Work

There are various methods available for the contractor to price his construction. The two most widely used in some form are the cost-plus approach and the lump sum method. The cost-plus approach (sometimes called the "labor and material") in its simplest form allows the contractor to total all actual costs for labor, materials, supplies and equipment (i.e., the actual costs of construction) and add to it a factor for his overhead and profit and pass it along to the owner. The contractor turns over his receipts and adds a percentage to the total, which can vary from as little as five percent to as much as thirty percent (or more). This money represents the contractor's payment for his firm's internal costs in building the project, as well as his profit. This form of pricing removes virtually all risk from the contractor and places it squarely on the owner, as the contractor is guaranteed payment of all his costs and his profit. This form of pricing is usually seen on small projects, of a size that will not sufficiently tempt the contractors to bid the project, or where there is insufficient time to do proper cost estimating (e.g., change order pricing).

Lump sum pricing allows the parties to rely on certain experiences, opportunities, and advantages that they may have in putting together a price that will entice the owner while allowing an appropriate profit by achieving such economies. Proper pricing depends upon the contractor's thorough knowledge of the scope of the work from the contract documents, the site characteristics, as well as the cost and availability of labor, materials, and reliable subcontractors. It also depends on appropriate scheduling so he can keep his crews working with little or no down time. The contractor is therefore able to take advantage of other opportunities immediately before or after the job on which he is currently working. The owner and architect need to be familiar with these factors and plan for them in the construction contract. Including provisions in the specifications that allow the contractor to seek legitimate change orders for unforeseen or hidden changes in conditions, or for cases in which the owner seeks additional work not legitimately called for in the contract documents, will give a contractor the confidence to bid his best price, knowing that he will not be harmed when circumstances beyond his control result in extra costs. Owners can "level the playing field" by further including clauses acknowledging the contractor's familiarity with the scope, the site, and the external labor and material market, and including provisions for *no damages for delays*, as well as liquidated damages for unexcused delays.

THE PROCESS

It all starts with an idea. An idea that occurs in the client's mind long before he calls the architect or the banker. The idea will have various features, components that are essential to the client and represent the core of why he is building. These features may range from the simple and spatial (larger showroom) to the esoteric and aesthetic (enhanced image, fresh appeal). It is important to remember this starting place, because at various points in the project questions and challenges will arise that will threaten these beginning goals. Scope, schedule, or budget will intrude on this initial vision and make demands that can erode the beginning ideals. Architects sometimes view the energies and talents they bring to interpreting their client's vision as the most important elements of a project. They are certainly important, but architects who forget that their client's needs and monies are the heart of any project do so at their own peril.

The process of creating a building can be distilled into three key steps (see Figure 1-2). The entire process is often envisioned as a funnel, wide at the top and narrow at the bottom. Each step represents a narrowing of the unknowns until the entire scope of the project is defined tightly and completely in the construction documents. Experienced architects know the process rarely works that cleanly. Owners think—and rethink—their decisions, sometimes circling around a problem and arriving back at an earlier spot. Some

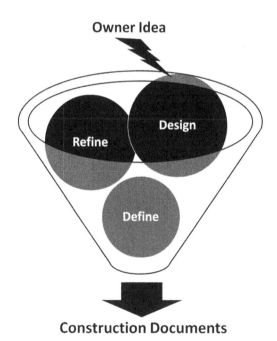

FIGURE 1-2
The design, refine, and define process.

clients are notorious for reactionary changes during creation of the construction documents, demanding that the plan or detailing be revised to accommodate new information, a consultant's recommendation, or a personal whim. Despite these occasional diversions and aggravations, the overall movement of the process should be from wide and exploratory to narrow and defined. That is the means through which quality documents are created. And make no mistake about it: High quality construction documents mean reduced risk.

Design

For his first step, the architect is normally responsible for defining the *program*, a written narrative statement of the owner's needs. The program should include as much detailed information as possible, but its main purpose is to describe the purposes of the spaces in the project and their relationship to one another. A program requirement for the sales floor of a small retail establishment might read: "Create a two-level sales area, with the upper level approximately 2 feet above the lower, and containing a feature slat wall illuminated by accent lighting, suitable for hanging art objects weighing no more than 50 pounds." This single sentence communicates a great deal of information about the sales area, and though much more information is required, the owner and architect have already begun to come to a shared understanding of the nature of the spaces the architect will design. A program, in this sense, is an invaluable tool for setting goals and expectations. Programs can be short and basic or elaborate and heavily detailed, depending on the nature of the project. Whatever the length, the program needs to have two characteristics: It fully describes the extent and goals of the project, and the architect and owner agree it is achievable.

The second step in the design phase is exactly that—designing. Typically referred to as schematic design and preliminary design in architectural agreements, these phases are actually distinct in large projects but are often combined in projects of smaller scope. Schematic design consists of creating two to three distinct alternative means to resolving the problem. This usually takes the form of differing floor plans at first, but can be expanded to include alternative exterior elevations as well. The purpose of presenting diverse schemes is to quickly flush out owner preferences. The presumption is that an owner has some pre-conceived preferences at the onset of a project. He may not recognize them until he sees a graphic representation of his space, but he inherently knows what he considers most desirable for his needs. By offering several distinct options, the architect forces this preference out into the open. This process is much more desirable than having the owner's vision evolve over a longer (and less profitable) series of drawing iterations as the owner guides and pushes the architect toward what he wants. Once again, the process rarely works this cleanly. Most often, the owner and architect agree on a preferred scheme that requires more refinement. Even if the owner does not find that one of the two or three initial schemes fully meets his needs, the time has not been wasted. At least the architect has identified what the owner *does not* like!

On larger projects, preliminary design is a distinct phase where the most desirable schematic design scheme is expanded and explored in more detail. Exterior elevations and preliminary sections are developed. Primary finishes are selected, at least generically, and the site design and structure of the building is determined. Preliminary design for office spaces or retail may also include a *test fit*, an insertion of seating, cubicles, rough furnishing layouts or other space planning components to determine if the preliminary plan will meet the owner's needs. Architects should view preliminary design as a way to gain the client's agreement on the principal design elements of the building. Of course, some of these pieces will change later, but hopefully for specific and well-reasoned causes. The architect's goal should always be to keep the design process moving from the general to the specific—and guide the owner in that direction.

One important tip: Give each phase, schematic and preliminary, a definite starting and stopping point. The architect should announce to the owner, "We're moving on to preliminary design," and let him know what the next phase will consist of and where it will end. Most importantly, have the owner sign the final drawing at the end of each phase. This signifies an end point for the phase, and at least a tacit agreement that the architect is now entitled to invoice in full for that phase and move on to the next phase of the project: Refining.

Refine

The *design development (DD)* phase of an architect's contract represents the period to take the design and refine the details of appearance, structure, mechanical, electrical, and plumbing systems; and to select finishes. Although the extent of documentation varies, the end point of DD is often defined as representing approximately 40 percent of the construction documents (or approximately 15 percent of the overall fee). By the end of design development, the architect—and owner—should be reviewing floor and reflected ceiling plans, building sections, exterior elevations, preliminary lighting and HVAC plans, and outline specifications. Design development also adds physical constraints to the design, such as overall building and material size, weights, heights, and depths. Because of the amount of detail being developed, the DD phase also offers the opportunity for a critical budget check. Using in-house resources, a construction manager or paid contractor, or an outside cost estimating consultant, the architect and owner can use the DD documents to obtain a fairly reliable cost estimate for the work. The benefit of obtaining this input can be invaluable. Because the architect is not so deep into the detailing of the documents, he has more opportunity at this point to realize cost savings through major changes in the building systems, structure, or skin. When handled well, DD accomplishes another important goal. It engages the owner more fully in the details of planning his facility. For the first time he must begin considering such aspects as maintenance and utility costs, building security, and telecommunication features. By the end of the design development phase, the owner should have no doubt as to what he is building. Many details still need to be worked out, but he knows and understands the project—and agrees in writing to the completion of the construction documents.

Define

The final phase is that of definition, taking the work produced in the earlier designing and refining phases, and defining it to such an extent that a contractor can build from the result. That is the simplest definition, because contract documents (as discussed more fully in later chapters) have other purposes as well:

• They are the permit documents: Code officials will review them to ensure that the building meets all applicable codes.

• They are the bid documents: Bidding contractors, subcontractors, and suppliers will scour them, do quantity takeoffs, analyze the specifications, and prepare bids based on what they find.

• They are the contract documents: Once a successful bidder is selected, the drawings and specifications become the key piece of the contract defining what the contractor is to provide for his bid.

• They are the basis for the owner's record set: At the end of the project, either the contractor or architect will update the drawings with as-built information for the owner's use in maintaining or altering his building in the future.

Of these uses, bidding of the construction documents is arguably the most demanding of the architect in terms of the scrutiny he will face from a group of bidders looking for flaws in his documents. This is also the phase in which the owner places the most reliance on the architect's professional expertise. Missing scope, unclear or insufficiently detailed work, contradictory information, incorrect dimensions or unbuildable details, uncoordinated systems, and other failings plague design professionals and vex owners. With the complexity of systems present in even a basic building, the potential for coordination errors resulting from a team of consultants preparing documents across three or more disciplines is always present. Recognizing that most of the risk to the architect comes from poorly prepared documents should actually be comforting. The one area in design that generates the most potential risk for the architect is the one where he has the most control and the most training to manage.

2

Agreements and Contracts

Agreements and contracts form the basis of what parties in a business relationship expect of one another. When everything is going well on a project and all parties are reasonably satisfied with the performance of the others, "what's on paper" is largely forgotten. But when questions are raised about competence or completeness, the agreements and contracts are scrutinized closely for what they say—and what they do not say.

Chapter 1 highlighted the importance of the contract in construction. The agreements too often do not get the attention they deserve at the time it matters most—before execution and the commencement of construction. For that reason, it is imperative that the architect make sure his own house is in order, in the form of an acceptable professional services agreement, before devoting his time and energy to the creative and collaborative process of design for his client. After bidding and negotiation, the architect will undoubtedly be called upon to assist the owner in formulating a construction contract. This will likely include the basic agreement, bid addenda, responses to bid questions, and other documents provided as part of the bid package. All of these together, along with the plans and technical specifications, comprise the contract documents. While the basics have been covered in Chapter 1, more specific suggestions on how to make the most of the contracts are offered here, first for the owner-architect agreement and then for the owner-contractor agreement.

THE OWNER-ARCHITECT AGREEMENT

Defining Scope

Architects base their fees on an estimate of how much time they expect to spend on designing the project and preparing the documents. Their fee is based on the scope of services the architect defines in his agreement. This time is almost entirely based on the scope of work for the project. The scope of work is simply the owner's end goal for the project. Like the plot of a novel, the basic scope should be easily expressed in a few words (such as a 14,000-square-foot single-story ophthalmology outpatient center). Having based his fee on certain assumptions regarding the scope of services he is to perform, it is critical for the architect to state these assumptions in his agreement with the owner. If the owner disagrees, he will certainly say so and the scope (and perhaps the fee as well) can be revised. A failure to clearly state the scope of the services in an agreement effectively denies the architect the ability to argue for extra compensation when the scope changes. The owner may or may not remember previous verbal agreements regarding scope. He may point to the contract and ask, "Where does it say you are not responsible for this?"

A quick word about the differences in scope terminology between design professionals and contractors is in order. *Scope of work* is the term normally used to define the construction extent documented in the construction documents by the architect—in other words, the construction work. *Scope of services*, on the other hand, is a term used to describe the range of architectural and consultant services provided to the owner to create the documents. The distinction is important, particularly in the language of contracts.

In defining the scope of services in his agreement, the architect should state efficiently and clearly what he understands the scope to include, and in equally short sentences, what it does not include. Here is an example:

> 1. *Design of a new single-story 12,000 square-foot retail building, with masonry load-bearing walls, steel joist roof framing, and exterior insulation finish system with built-up parapet on*

Risk Reduction Tools

Recommended Owner-Architect Contract Provisions

- Define owner expectations clearly.
- When is the contract in default?
- How much and how often will the architect be paid?
- How will the owner and architect resolve disputes?
- No verbal agreements—Everything must be in writing.
- The owner cannot sue for anything insurance covers.
- Statute of limitations period begins with substantial completion.
- Limit or lessen liability as much as possible.
- Liability reduction techniques include: limiting clauses, indemnity, additional insured provisions, and subrogation waiver clauses.

the front and one side. Building to be subdivided for four tenants, separated by steel stud partitions;

2. *Scope to include the design of new rooftop heating, ventilation and air-conditioning, electrical power and lighting, and code-minimum toilet fixtures. Fire protection scope to be limited to performance specifications by the engineer, with the fire protection subcontractor responsible for final engineering;*

3. *The scope does not include renovation to any part of the existing center. The scope does not include any tenant fit-outs or interior finishes beyond those associated with the toilets.*

In a relatively brief statement, the architect has defined the scope of what he will and will not do on a project of limited size. Even on more complex projects, however, the statements describing the scope can be almost as terse. For complex projects, or those where the fee is based on a number of assumptions, the architect can attach a separate program document stating the assumptions. He will always have the option of performing minor changes in the scope for no additional fee if he wishes, but without definition of the scope in the agreement he has little basis on which to request an additional fee when the project experiences what architects euphemistically refer to as *scope creep*.

Indemnification—A Trap for the Unwary

Often the owner will want to shift some of the risks inherent in the construction process, such as personal injury, property damage, and economic loss due to delays (consequential damages), to the other construction participants—including the architect. "After all," reasons the owner, "I am paying a lot of money for services, and since I am neither designing nor building the project, why should anything that goes wrong be my financial responsibility?" This sentiment sees its most extreme manifestation in the *indemnity clause*, which expresses the unrealistic desire of the owner to immunize himself from all risks while still receiving all the benefits of project ownership. Indemnification language often appears as *boilerplate* (repetitive, standard language for recurring situations) in agreements. It is included in this manner to lend it credibility and the appearance of immutability. Indemnification clauses typically require the architect to stand in the shoes of the owner in the event anything goes wrong during the project. This includes those instances where there is damage or injury to a third party with potential financial consequences to the owner, including the attorneys' fees necessary for defense of the matter. These clauses are widespread and inescapable. The architect, especially one without bargaining leverage, can only hope to limit the impact of such a clause with one or more of the following strategies:

1. **Uninsurability**. The clause should only allow indemnity for *the negligent acts or omissions* of the architect. The owner should not be indemnified merely in the event that "something goes wrong" that is not the fault of the architect. This type of "any and all" indemnity is referred to as *broad form*, and is often not covered by an architect's professional liability insurance carrier. The owner should be warned about the risk to himself of invalidating the architect's coverage for this type of indemnity.

2. **Public Policy.** The state in which the project is located may in fact forbid or dramatically limit the use of indemnities of either the broad or narrow form. If state law forbids them, indemnities should certainly not be included by the architect or owner as their inclusion may be seen as waivers. It is better not to include an indemnity clause than to ignore it in the contract in the belief that it will be ultimately voided in court as against public policy.

3. **Reciprocity.** If the owner insists on indemnification, so should the architect. By asking for mutuality of indemnity, the architect essentially obtains protection roughly equal to what he gives up. Faced with the choice of indemnifying the architect, the owner may drop the subject altogether. Where possible the language of both the owner's and the architect's indemnification should match.

Limitation of Liability

The architect may seek to protect himself from unacceptable or unfair risks through the use of a *limitation of liability* clause. This clause typically limits, but does not completely eradicate, the architect's financial risk during and after construction to the owner. Unfortunately it is virtually impossible to limit the architect's risk to third parties with whom the architect does not have a contract. Simply stated, the architect has no contract with them and therefore cannot eliminate his duty to refrain from negligent conduct that runs a foreseeable risk of harm to those who live or work in the building. Typical liability limitation clauses can limit the architect's financial risk to the owner to a lump sum amount, or to the architect's fee, or the greater or smaller of either of these amounts. Note that while some states disallow any attempt of any professional to limit or eliminate risk for professional services, most states allow limitations within reason as long as the provision does not attempt to eradicate the risk altogether. Even limiting the architect's risk to no more than the amount of his available insurance coverage (either professional liability or general liability) provides some benefits as the owner's leverage of seeking more than the amount of the architect's insurance coverage—an especially frightening prospect to most architects—is thereby removed.

More comprehensive contracts, especially in large projects, allow limitations where architects need it most—in change orders. The architect may be able to negotiate the limitation of liability for so called *error and omission (E&O)* damages. These are defined as changes to the scope with financial consequences, and additional fees to the contractor resulting from errors or omissions in the contract documents. The owner may be willing to waive such damages that result from an agreed upon, though arbitrary, percentage of increased construction costs due to such errors and omissions problems. For example, the owner may be willing to waive damages associated with the accrual of E&O changes not exceeding three percent of the original construction price. Of course, not included in the calculation of this amount are change orders resulting from an increase in scope, differing or unforeseen site conditions encountered by the contractor, or the owner's discretionary or voluntary decision to pay a requested order, arguably part of the contractor's basic scope.

Another way to limit liability is by time. Most states have statutes of limitations that fix the amount of time within which claims must be filed, after which the right to claim damages is lost forever. However, the law of contract in most states also permits the parties to modify the law for claims brought under the contract. For instance, the parties to an agreement can agree to a shorter or longer time period if they believe it suits their particular needs. Some states have a four-year statute of limitations, which usually begins on the day one of the parties believes they have been damaged by the other party's breach of the contract. The parties to the architect/owner agreement may agree, nonetheless, to require a filing of all claims within two years (or any other mutually agreeable length of time) from the date of substantial completion. By agreeing to this, not only have the parties shortened the period of limitations but they have further truncated the period by making it commence from a certain date. This date is commonly set as substantial completion, rather than the "moveable feast" date known as the date of the injury or damage that is used in most statutes. Beware, however, because the limitations period is usually reciprocal, which means that what applies to the owner also applies to the architect as well. But since the architect will usually have less cause to file a claim against the owner, he should benefit more from the shortened time. In any event, when negotiating limitations of liability, either as to substantive acts or time, the architect should seek the advice of legal counsel.

Certifications and Assurances

Often the architect is required to provide information or opinions, sometimes in the form of legally binding language, about the quality or quantity of construction. The architect should look first to his agreement to determine if such service is contractually required and, if it is not, he should think twice about signing a document that renders him potentially liable to another based on the architect's certification. The circumstances most likely to increase the risk to the architect occur when the owner's lender requires regular updates on the progress of construction. While the architect should anticipate providing information reasonably based on his observations of the quantity of construction, he should avoid assertions about the *quality* of the construction. It is unlikely that the architect's contractually-required observation duties are sufficiently comprehensive to prepare him to discuss quality of construction issues, especially as they relate to the contractor's workmanship. Any contractual obligation to provide such certifications should be limited to the scope of the architect's existing duties of observation (i.e. visual only, no extensive, invasive, or destructive testing or inspections) and further provide him with enough time to review and complete the form required. This review period should be set at a minimum of 14 days.

Exclusions and Additional Services

Additional services are an area rife with disagreement between the architect and his client. The best way to avoid such disagreements is in the first stage of the relationship—the agreement. By carefully defining services that are included—and excluded—in the architect's agreement, many of the common disputes can be avoided. In listing exclusions, the architect must be careful not to exclude any services that are reasonably required or expected by the owner. If, for instance, the owner has made it clear he expects the architect to include an outside cost estimator as part of his scope of services, the owner will be understandably annoyed to later discover the architect has considered it an additional service.

More often, disputes over additional services occur over plan changes or scope alterations late in the construction documents phase. Owners react to advice from a number of sources, and a tenant or vendor (or even a spouse) can persuade him to modify his plans at any point in the project. The potential for disagreement can be alleviated if owners are willing to *sign off on the plans*, or verify agreement with the documents at various stages in the project. Owners are notoriously reluctant to sign off on drawings at various stages of the project. They recognize that it is the architect's attempt to limit their ability to make future changes without paying additional fees to the architect, and architects rarely press the case when owners refuse to sign off. The best solution is to show the owner the consequences of his change, detailing the sheets and disciplines affected, and the estimated time necessary before providing the revised services. Given this information, owners sometimes decide the change is not worth the cost.

A third area where additional service disputes occur is when bids are returned that exceed the owner's budget. According to the standard AIA agreements (AIA Document B141©), if the owner clearly conveyed his budget requirements to the architect and cooperated in adjusting project scope and quality if necessary, then he may be justified in asking the architect to revise the documents and rebid the project at no additional cost. Where this proposition becomes dicey is in those projects, such as prototype retail spaces or hotels, where the scope is tightly defined and the budget is constricted. Architects have little room to maneuver with prototype designs, and much of the cost may have been determined with standard details provided to them by other design professionals or the owner's own consultants. An architect who is aware of these situations ahead of time should discuss with the owner how bids that exceed the budget will be fairly handled. Will the owner pay the architect to revise the scope (and offer guidance on how to do so), or will the owner increase his budget?

Termination or Suspension of Services

There should be no question as to when and under what circumstances the parties have the right to suspend or terminate services and, if necessary, sue for breach of contract. This includes establishing the acts or omissions of either party that constitute a breach. Here it is perfectly acceptable to state the obvious, such as the fact that the owner's failure to pay invoices within a certain time period constitutes a material breach of the owner's duty and permits the architect to suspend services until payments are made current, without responsibility for any delays that occur to the project due to the suspension.

A failure to express such a sentiment may obligate the architect to continue in the owner's service to its completion and only file his claim at project's end. Not only does the architect lose considerable leverage in such instances, he runs an excellent risk that the project will run out of money while he is still hard at work. In the event of continual or habitual lateness by the owner in making payments, the architect may wish to consider terminating the contract altogether. Either way, professional responsibility requires that he give the owner a reasonable opportunity, in writing, to cure the breach prior to suspension or termination of the contract. Seven to 14 days should be sufficient for this purpose. Indeed, it would be wise to require both parties to give written notice of their intention to suspend or terminate, preceded by an opportunity to cure the alleged defect in performance.

Project Schedule

A useful tool the architect can employ in helping the owner to understand the amount of time necessary to carry a project from idea to completion, and to realize the urgency of decision-making on his part, is to create a project schedule. A project schedule should include general assessments for each major task or milestone in the project, including planning and zoning approvals, phases of architectural services, bidding and negotiation, and construction. See Figure 2-1 for an example of a basic project schedule.

Construction Contract Administration

The duties of the architect on the project, after design and the owner's selection of a contractor, are often referred to as *construction*

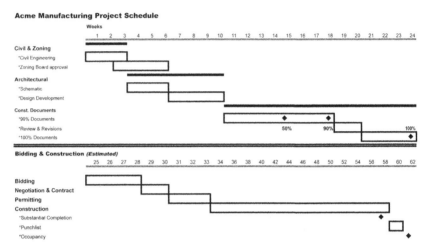

FIGURE 2-1
Sample project schedule.

contract administration or *construction administration (CA)*. This is defined simply as the process of assisting the owner during construction. Construction administration can include as little as being available to answer questions about the construction documents to as much as complete management of the construction project, including (in special cases) supervision of the contractor's performance. The key element in either extreme is the absolute requirement that the owner and architect agree on the description, nature, and type of CA services the architect will and will not provide on the project. However, construction administration typically requires the architect to provide a series of services which form the bridge between the delivery of the construction documents and the contractor's performance of the work. See Figure 2-2 for a list of commonly provided construction administration services. These tasks are related to the process of refining and explaining the design, and confirming that the contractor's work is generally in conformance with the intent of the construction documents. These services include preliminary communication with the contractor, submittal review, responding to RFIs, evaluating substitution requests by the contractor, observing the work for conformance to the construction documents, review and approval of payment requests, review of the contractor's request for change orders, and assisting the owner in approving or rejecting them. The architect is usually charged with the responsibility of determining *substantial completion* (the point at which the owner is able to use the work for its intended purpose) or *final completion* (the completion of all punchlist work, the turnover of all warranties and forms, and completion of all close-out processes). Beyond project representation on behalf of the owner, the architect may also be called upon to play the role of arbiter between the owner and contractor in matters of interpretation of the intent of the construction documents. In such instances (which should be spelled out clearly in the contract) the architect is no longer functioning as the owner's project representative but as an unbiased, nonpartisan decider whose primary allegiance is to

Construction Administration:
The Bridge from Construction Documents to Completion

❏ Pre-Bid Conference
❏ Bid Requests for Information
❏ Owner/Contractor agreement review
❏ Construction Project meetings
❏ Periodic Observations and Field Reports
❏ Submittals
❏ Substitution Requests
❏ Change Order Requests
❏ Construction Requests for Information
❏ Punchlist
❏ Substantial Completion
❏ Closeout Document Review
❏ Final Completion

FIGURE 2-2
The construction administration bridge.

the construction documents and the design intent. In such instances, the architect should require immunity for his services rendered in good faith, from both the contractor and owner in the owner-architect agreement and in the construction documents.

All of the duties mentioned above are significant to the project and are often misunderstood by the public consumers of architectural CA services, mainly owners and contractors. For this reason it is worth the time and effort (and space in the contract) to call out the specific services that will and will not be provided. It is equally important to record the architect's obligations (attorneys call it *memorializing*) in the agreement. It also can smooth the path, especially for the first-time owner, to review the tasks that the architect will perform, a brief description of why and how the architect accomplishes these tasks, and the common obstacles and stumbling blocks that occur. It is even more valuable for the architect to state in the contract how most impediments occur, and the basic methods the architect will use in dealing with them. The contract thereafter becomes a testament to the reasonable expectations of both parties on the path to a completed project.

Construction Costs

It is commonly understood that cost drives construction. Other than the idea of the construction itself, cost issues dominate the process. Owners are constantly seeking to remove or reduce uncertainty in the costing process to help them predict the ability to start and finish the project, and make a profit at it regardless of their use or intent. Unless the owner is very sophisticated and experienced, his first source of information on the cost of construction is the architect. The architect certainly has the ability to compare the owner's budget with known estimates for the cost of construction, or even to seek out guidance from other areas to help assess the cost of the work. In providing cost estimating advice, however, the architect can do little more than use his best judgment as a design professional familiar with the construction means available and the relative state of the market for labor and materials. This recognition must be reflected in the contract or the architect may find that the owner considers him an expert in cost control, and that he should therefore be held responsible for bids that exceed the owner's budget. There are several key elements to limiting the architect's liability for a runaway bid.

1. *Limitations of Estimate:* Make sure the contract reflects the inability of the architect to predict construction costs, especially where the market is volatile and there are few resources available to help in this task. The architect may wish to suggest to the owner the use of outside professional or sub-contracted construction cost estimating services to provide him with independent and more capable budgeting services. If the owner declines, the architect should document his unwillingness to go this route.

2. *Project Contingency:* Include a project contingency factor for the bid process. This *flex* in the estimate allows for some line item estimating errors or misjudgments to be absorbed in the overall project budget.

3. *Bid Time Limits:* Place realistic time limits on the bid so the owner's delay in procuring bids does not wind up being the reason for significant increases in projected costs.

4. *Fast-track and Phasing:* When construction costs are rising rapidly, suggest fast-track construction or phasing to the owner as a way to get earlier and more favorable bids locked in to protect against cost increases. This requires coordination of the documents to ensure that items such as structural steel and HVAC equipment can be safely purchased early. Notify the owner in writing of this suggestion, and document if he rejects this proposal. Additionally, allow all bidding (except where prohibited in public sector bidding) to include a negotiation of the price. Finally, the architect who has a client with an extremely tight budget should consider expressly limiting his responsibility to the owner to correct the consequences of high bids. This may consist of stating that a single redesign to meet the budget is the sole remedy for a situation where all the bids from contractors exceed the owner's budget.

5. *Insufficient Budget:* If the architect is convinced that the owner's budget will not support the work, he must advise the owner to either increase the budget or reduce the scope.

While the owner will not be happy with either option, it is critical for the architect to advise him of this problem. If the owner decides to proceed with the original scope through bidding to "test the market," the architect must make clear that he is entitled to additional services for any redesign services if the bids come in higher than the owner's budget.

See the Risk Reduction Tools for a checklist of contract hazards.

Deliverables

Architects produce services that are sometimes unfairly discounted by the owner. The end result of months of work and tens or hundreds of thousands of dollars in fees is a roll of prints and a thick book of specifications. A whole world of information is included in the documents, but on the surface they are not overly impressive to an owner. He sees them as a tool to an end. If he comes to believe that tool is badly flawed, the owner may feel that he has recourse to hold remaining payments from the architect or pursue in the courts a refund of the fees he paid. At that point, the question of deliverables comes into play: *What did the architect owe the owner, and did he provide it?*

 **Risk Reduction Tools**

Contract Liability Hazards
- Unclear or ambiguous scope
- No mention of exclusions or additional services
- Onerous indemnification clause
- No subrogation clause
- Starting work without an agreement
- Owner demand for certifications or assurances
- Inadequate or unclear budget
- Owner insistence on scope and quality for fixed budget

Field Observations

Field observations are a defined part of an architect's service to the owner. They are described in such general terms that an outside observer might wonder what value this service truly offers the owner. Field observation: *The periodic visit to the work site by the architect for the purpose of generally observing the progress of the work.* The general language offers both protection and challenges to the architect. On the one hand, he is not required to verify that the contractor is conforming to every aspect of the contract for construction. The contractor is responsible for meeting his contract. The architect, on infrequent visits to the site, is responsible for generally observing the progress of the contractor and, on behalf of the owner, alerting him to any concerns he has regarding the pace, contract conformance, or quality of the work. A contractor disposed to conceal deficient work from the architect can easily do so. If an owner has cause to be concerned in this area he should not look to the architect for the solution. Few architects can maintain staff full-time on a construction site (though some owners do pay for such service). The owner can hire his own representative (a so-called *clerk of the works*) to represent his interests at all times on the construction site. This may be valuable on a large and complex project with many trades, but is not necessary on typical projects. Most contractors value their reputation too much to cheat an owner, or to allow their subcontractors to do so, on any given project.

One aspect of field observation in which architects are under increasing pressure to be diligent is work site safety. Safety on a construction site is, by contract, the sole responsibility of the contractor. Recent court decisions, however, have held the architect and his consultants responsible for alerting contractors and owners to unsafe conditions they actually observe on a work site. If he observes an unsafe condition, the architect must take immediate action to notify the contractor of the condition and insist that it be corrected. If the contractor refuses, the architect should notify the owner of the situation and insist that he authorize suspending work on the project until the unsafe situation is corrected.

At each field visit, the architect should document the progress of the work with a field report and photographs. The field report form can be a single-page document that lists the date, weather conditions, the number and type of personnel observed on the site, the work they were performing, and the general progress of the work. If any special questions or conditions arise during his field visit, the architect should record them on the field report as well. The field report is generally distributed to the owner and contractor. Since field visits are often conducted in conjunction with project meetings, it is important to note that the two are essentially different activities. The project meeting will cover in-depth issues of the project, and the results of those discussions will be recorded in meeting minutes. The field report is a general observation and snapshot of a day in the life of the project, and the nature of the report should reflect that. The architect will certainly list a few areas of concern discussed in the meeting, and these concerns should be reflected in the field report and photographs. Any decisions or assessments made on particular issues should be the subject of separate correspondence, however, and not shoe-horned into the field report for convenience. On the other hand, general observations of poor workmanship observed during the visit should be recorded on the field report and backed up with a photograph.

Alternate Dispute Resolution

Standard agreements between the owner and architect (as well as the owner and contractor) usually list arbitration as the preferred means of resolving otherwise irreconcilable disputes between parties. Arbitration consists of a semi-formal hearing conducted by a neutral party experienced in the areas of design and construction. Arbitration can be binding or non-binding. In binding arbitration, both parties agree to forego the right to sue the other party and accept the arbitrator's decision as final. Both parties agree on the arbitrator(s) and the date of the hearing, and submit all relevant documents in advance of the hearing for review by the arbitrator. Attorneys are permitted but not required. In arbitration, excessive legal maneuvering is not really helpful to a claimant's cause. Arbitrators are trained by the American Arbitration Association (AAA) in how to conduct hearings and interpret the evidence, which is always aimed squarely at contractual agreements and the reasonable obligations resulting from them. Arbitrators have been criticized for too often "splitting the difference," and burdening each side with roughly half the cost of the dispute. Statistics developed by the AAA show this is not really the case, with arbitrators most often choosing a clear winner in a dispute. Arbitration does, however, suffer one serious shortcoming in construction. It is slow and costly. Arbitration hearings can take months to arrange, and can cost each party thousands of dollars in fees for the arbitration, documents, and legal counsel. Meanwhile, the core issue that prompted the arbitration has probably been dealt with out of pure necessity: the contractor has performed the work in question without a change order, the owner has paid a change order he feels the architect or contractor is responsible for, or the architect has relented and provided additional services without compensation. In the interest of keeping the project moving, one of the three parties in the triangular relationship has acted to resolve the problem, and absorbed the cost of doing so.

Another method widely used, as either a replacement for arbitration or as a predecessor to it, is mediation. In this method of dispute resolution, a trained mediator is brought in to help both parties in a dispute to come to agreement on an equitable resolution. This method is non-binding and is usually faster than arbitration. Its weaknesses are that it relies on good will and flexibility on the part of both parties, and by its nature tends to move the dispute toward middle ground where each party gives a little and gets a little. See Chapter 7 for a more thorough review of dispute resolution techniques and options.

THE OWNER-CONTRACTOR AGREEMENT

The agreement between the owner and the contractor is of particular interest to the architect, who should review it carefully because its contents include information relating directly to his construction administration services. The risks the architect assumes during construction will depend in large measure on the content of that document. The owner-contractor agreement can take various forms. All such contracts normally describe in detail the project and the price, including any method, conditions, and limitations on payment as well as all necessary and essential terms. To avoid the document becoming too long or cumbersome, especially in large projects, the parties often incorporate by reference other documents that they want to make part of the agreement. Such incorporated documents should always include the plans and specifications/project manuals; instructions to bidders; any addendums; the architect's supplemental instructions; and all approved sketches, modifications, and changes to the contract during negotiation. By incorporating these documents by reference, the owner makes a failure to follow these design documents a breach of the contract, rendering the contractor liable for the contract remedies of damages for breach, back charges for owner-completed work, liquidated damages and, perhaps the contractor's biggest fear, withholding payment for incomplete or out-of-specification work.

The architect should consider, at the very least, using the division one specifications offered by the Construction Specifications Institute or other boilerplate general conditions such as the *AIA-A201 Standard Conditions of the Construction Contract*© as a starting point for developing the terms and conditions appropriate for his particular needs. Fully analyzing the anatomy (and pathology) of the construction contract is the stuff of law school theses, but for the architect it is important to examine some basic provisions that should be included in the terms and conditions of the construction contract.

Insurances and Indemnities

All contracts should require the contractor to carry the full range of insurances including: general liability (for personal injury or property damage in the event of the negligence of the contractor); motor vehicle insurance (liability for vehicular-related accidents, usually excluded from general liability), workers' compensation (coverage usually required by state law for on-the-job injuries to the contractor's employees but not immunizing the owner from suits for negligence due to the contractor's employees) and builder's risk insurance (property loss during the project prior to completion). Proper coverage amounts should be required by the owner after consulting his own insurance broker on the subject. The owner should also require current certificates of insurance from the contractor's broker to prove the existence, location, and amounts of coverage (with an appropriately rated insurance company) as proof of adherence to the contract requirements. See the Risk Reduction Tools for a checklist of insurance types.

Wherever possible, the contract should also require that the owner, the architect, and all owner's agents and representatives be named as *additional insureds* under the policies. This means that the contractor's insurance company names the owner and the architect as persons who are insured under the policy to the same extent as the contractor. Special endorsements at little or no additional cost are available to the contractor from his insurance carrier. The contract should include a requirement that the contractor provide proof of existence for all the required insurance coverage before the commencement of any construction. Although not strictly part of the architect's scope, if the architect becomes aware that the contractor has begun work without the contractually required type or amount of insurance (or without proper proof that the contractor has named the owner or architect as additional insureds), he must advise the owner immediately of this fact, along with the options permitted by contract. Such options should include stopping the work (to avoid at-risk, uninsured work), the purchase and back charge of necessary coverage (if provided in the contract), or even terminating the contractor for refusal to purchase the required insurance.

 *Risk Reduction Tools*

Contract Insurance Checklist

1. General Liability Insurance
2. Motor Vehicle Insurance
3. Worker's Compensation Insurance
4. Builder's Risk Insurance
5. Payment or Labor & Material Bonds (if required)
6. Coordinate required coverage with owner
7. Obtain insurance certificates from broker
8. List architect and owner as additional insureds

While not strictly an insurance policy, bond (suretyship) requirements are also an appropriate subject of the contract. The often
prohibitive cost and conditions of surety bonds for performance
or for labor and materials—costs which are always passed down
to the owner—make them unlikely on all but public-sector jobs.
If bonding is required by the owner, specific provisions should be
made for the amount, the ratings of the company providing the
bond, and agreement for an understanding of the conditions under
which the surety will step in to take over responsibility for the construction. Copies of the policies need to be provided to the owner
and the architect. The owner will normally look to the architect to
advise him on the necessity of filing a claim under the suretyship
agreements. The architect must, however, avoid giving legal advice
to the owner and stick to the factual issues, such as the status and
progress of the work.

Just as the owner will seek indemnity from the architect, he should
certainly seek indemnity from the contractor as well. Because the
contractor is ultimately producing the total project for the owner at
considerable expense, and because the contractor is so thoroughly
responsible for the work site and all activities on it, there is a higher
justification for seeking indemnity of a comprehensive and broad
form variety. Additionally, the contractor can diffuse the risk more
easily among the various subcontractors, vendors, and suppliers
with written indemnities and additional insured clauses of his own.
The front-end documents must make very clear that the owner's
expectations are that the contractor will indemnify, hold harmless,
and provide the costs of a legal defense to the owner, the architect,
his consultants, and other direct agents of the owner as long as
the injury or damages arise out of the work, regardless of whether
the damages are the fault of the contractor. Anything unclear about
this provision will invalidate it for all claims and make it applicable
only to those claims that are the fault (by act or omission) of the
contractor or his forces.

Third-Party Relationships

Contractors need subcontractors to do the work of their trade or
specialty. Subcontractors need the contractors to pay them money so

they can pay their workers and suppliers, cover their overhead, and make sufficient profit to remain in business. These needs are often mutually dependent, and therefore mutually thwarted by construction disputes. The owner must make all legitimate attempts under the contract to limit the infighting and disputes that occur between a contractor and his subs, and to prevent them from interfering with increasing his project costs or completing the project within his schedule. To accomplish this, the owner must make it clear that the firms providing goods and services to the contractor are not third-party beneficiaries of the owner-contractor agreement. This will protect the owner from direct suits by the subcontractors for breach of the contractor's obligation to pay under a claim that the owner benefitted from the work of the sub. Without this contractual statement, the subcontractors will attempt to collect from the owner directly if the contractor defaults. The agreement must recognize, however, that the contractor has a duty to make all requirements between owner and contractor applicable to the contractor-subcontractor agreements so there is no gap in responsibilities.

Ambiguities and Inconsistencies

The preparation and coordination of architectural and engineering plans and specifications in large projects can be a difficult task. Notwithstanding sincere attempts at quality control, the possibility exists that the plans may not always agree with the specifications and that each of these parts may contain inconsistencies. Also it is possible that there may be an *ambiguity* in the documents. An ambiguity is an instance in the documents where reasonable minds might differ over the interpretation of contract language, particularly regarding the level of quantity or quality of contract requirements. For this reason, and to curtail the natural tendency of the contractor to seize upon, for his benefit, a mistake in detailing that could have been caught at an early stage, the front-end documents typically include language that resolves inconsistencies and ambiguities in the contract documents. While there are a number of ways to handle the problem, the consensus among case law decisions seems to be that the documents should express the requirement that the contractor, in bidding the project, must always assume that any ambiguity he encounters should be bid according to the higher quantity or better quality of building materials or systems resulting from the ambiguity or inconsistency. This places the risk of additional costs on the contractor if he has not read the documents carefully, is not fully familiar with the requirements of the project, or fails to ask pertinent questions before submitting a bid.

SPECIFICATIONS, SUBMITTALS, AND SUBSTITUTIONS

In order for the owner and the architect to satisfy themselves that the contractor knows and understands the nature of the work required by the contract documents, the specifications should require some form of submittal of written information that demonstrates that the contractor can supply, furnish, fabricate, and/or install the proper quality and quantities required by the design as expressed in the contract documents. The architect's review of these documents, however, is only meant as a check to see, generally, if the proposed work of the contractor conforms to the design intent. This is because the architect is typically not competent to address the workmanship of the contractor. After all, the architect is an expert in design, not in construction. Construction is the sole domain of the contractor.

Specifications should provide adequate information to the contractor as to the nature of these reviews for several reasons. First, the contractor should not send the architect late or haphazard submittals, since the schedule of submittals is tied to the schedule of the work. Job sequencing, the schedule of the rough versus the finish trades, and the ordering of long-lead items (those construction items requiring a great deal of time to obtain) all have to be coordinated by the contractor with an eye to the orderly and organized approval of the shop drawings.

Secondly, the contractor must understand that the architect's review is not for the purpose of commenting on the contractor's workmanship or directing the work of the contractor. The contractor is always responsible for the means and methods of construction and the architect has no right (and no duty) to interfere with the means and methods of construction. The contractor should not be in the position of blaming the architect or owner because he has made mistakes in the submittals that were not directly related to the design intent and not picked up by the architect. Neither dimensions and quantities, nor instructions for installation or performance of equipment installed by the contractor are proper subjects of submittal review and the contractor must be so advised in the specifications.

Finally, the documents must make clear that the contractor may not use the submittal process to substitute out-of-specification quantities, materials, or building systems. A recurring problem on projects is where the contractor substitutes a specified product or material, and after installing same without having received a valid substitution order from the architect, is told to remove and replace with the proper materials. The contractor often will refer to an approved shop drawing where a vague reference is made to the material, and the contractor will claim that the shop drawing in its entirety was approved by the architect. Only a comprehensive system for the approval of a substitution request coupled with an express prohibition of using the submittal process as a surreptitious substitution device will adequately protect the owner from this ploy.

TIMELINESS AND SCHEDULING ISSUES

The contractor is responsible for developing and managing his construction schedule. The architect has two roles on behalf of the owner: 1) Reviewing the initial schedule, and 2) Periodically reviewing schedule conformance with the contractor. In neither role does the architect have the right to instruct the contractor to alter his schedule. He may advise and express concerns, but the schedule always remains under the control of the contractor.

In reviewing the schedule on behalf of the owner, the architect is responsible for identifying those areas where he does not believe the owner's interests are well served by the schedule. In the initial review, these areas could include:

1. Insufficient time between phases to allow the owner to fulfill his obligations.

2. Completion dates for portions of the work that will impede the owner's operations.

3. Substantial or final completion dates that do not conform with dates stipulated in the owner-contractor agreement.

The architect should convey concerns in any of these areas to both the owner and contractor so the schedule can be adjusted to accommodate each of their needs. One problem that does occur during scheduling for work in an already operating facility is the contractor failing to realize that he will lose unexpected days in phasing the work, or will be working around the owner's need to sustain his operations. If this was not made clear, or at least not clearly implied in the bid documents, the contractor may request additional time and general conditions compensation for accommodating the owner's needs. This is a difficult way to begin a project. See Chapter 3, *Project Scope*, for information on the importance of including phasing information, however speculative, in the bid documents.

During construction, the architect and contractor should review schedule conformance at each project meeting. It is common for portions of the work to vary from the project schedule. Subcontractors can appear late or not fully staff the work. Complications arise that take time to resolve before the work can proceed. Code inspections can be delayed, or work that has failed inspection must be corrected and re-inspected. All of these occurrences are common and result in periodic fluctuations in the construction schedule. The contractor is usually well aware of why he is behind in the schedule, and should have a plan to recoup the time. But when the architect finds, over the course of several project meetings, that the contractor is chronically behind schedule and appears to have no well-defined plan to make up the lost time, he should report this to the owner and express concern to both the contractor and owner that the project completion date is in jeopardy. If the owner agrees, the project completion date can be extended and the schedule modified to reflect the new reality. In these instances, where the schedule deficiency is not due to any fault of the owner or unforeseeable outside factors, the contractor should not be entitled to any additional general conditions compensation for the extra time.

The architect also needs to be attuned to schedule issues on projects in which the contractor is working under a particularly tight schedule, or in which the contract contains a clause that obligates him to pay the owner a daily amount as liquidated damages if the contractor exceeds an agreed-upon deadline.

SCHEDULE OF VALUES

The schedule of values is the contractor's apportioning of the value of the work across construction divisions. Typically, contractors use the Construction Specifications Institute's *MasterFormat*© system to divide up the work into standard areas tied to subcontracts that can be efficiently invoiced and whose progress can be easily judged in the field. In general, the architect should check the schedule of values for these conditions:

- Is the full value of the contract reflected in the total of the schedule?

- Is the schedule of values excessively front-end loaded? In other words, is the contractor attempting to shift an excessive amount of the value of the work to early stages of the project? *(Note: Some front-end loading is common in projects. The question is one of degree).*

- Are there any major components of the work not reflected in the schedule of values?

- Are there any major components of the work that need to be broken into finer divisions to make field assessment of progress easier?

The schedule of values requires careful attention by the architect because, once approved, he will be living with it for the entire construction period. A schedule that does not fairly represent all the components of work can result in situations where the architect and contractor are negotiating where and how much of a hidden piece of the construction (one not represented in the schedule) should be allocated to another line item. Such assessments can get messy in a hurry, and leave both architect and contractor open to question if the schedule is ever scrutinized by outside parties. The most risk-laden situation of all is when the architect knowingly approves a schedule of values that is heavily front-end loaded in a misguided attempt to provide an underfinanced contractor with early working capital. This situation is rife with danger for the design professional in that it places him in a position of working against the fiduciary interests of his client and compensating the contractor beyond the fair value of the work in place. It is not hard to see how this situation can have catastrophic consequences should the contractor default.

COMMUNICATIONS AND NOTIFICATIONS

Lines of communications are important in any human endeavor and especially so in one as complex as construction. Therefore it is important to set out early the method and manner of acceptable communications. Where the architect is the owner's project representative, the contractor should relay all his communications through the architect. The architect will communicate with the owner and obtain authority to modify the contract requirements as necessary to secure a timely completion of the project. Conversely, neither the owner nor the contractor should communicate directly with the contractor's subcontractors or vendors except in well defined circumstances and, where possible, only with the contractor's consent. This helps to avoid charges by the contractor of interference with the contractual relationship between him and his subs. Also, certain types of damages may be alleged where the owner actively interferes with the work of the project. Such interference can certainly include contact with the subcontractors and vendors, especially where the owner or architect have given them direction independent of the contractor. While telephone communication is appropriate for its speed, most communications should be written down in a way that can be preserved as part of the project record. Some projects provide forms for the documenting of telephone calls, for sending and responding to requests for information, for the daily conditions reports, and for the minutes of job meetings.

When official communications are required, the contract usually sets forth the name and address, and occasionally stipulates the manner of communication, of the person who is to receive the notice for each party to the agreement. These provisions should be updated as required during the course of construction.

3

Construction Documents

DO YOUR RESEARCH

Architects are trained to work from the general to the specific. They are taught to gather basic information about their client's needs, project resources, and restrictions, and move toward a solution in ever-tightening circles of detail. This time-tested process works. What seems to have been forgotten in the rush of busy offices and tight schedules is the importance of understanding that a successful project occurs when the professionals involved focus on two simple aspects: 1) The resources, and 2) The goals. There are innumerable ways for architects to get on the wrong side of clients, but perhaps the easiest is for the client to realize that the architect has forgotten why he was commissioned in the first place. This section is an argument for spending some time at the beginning of a project to lay the groundwork—a foundation if you will—for what the end goal is and what resources are available to get there (see the Risk Reduction Tools).

This may all sound too easy. Computer-aided drafting drives architects toward specificity and detailing early. The process used to begin with fuzzy pencil lines on canary paper. Now it often begins with absolute dimensions and hard lines. Architects pull from standard detail libraries and often begin building the set of construction documents from the details. CAD-based plans are a wonderful tool that has enabled architects to work faster, and with better precision and coordination than ever before. But what can get lost in

Risk Reduction Tools

The Project Foundation

The Goals

- Make a list of the top five goals for the project
- Prioritize the list
- Create the architect's project plan

The Resources

- Know the owner's budget
- Create a preliminary project schedule
- Know the zoning and site restrictions
- Perform an early code check

the early attention to specifics is a fundamental awareness of the general: the limitations and characteristics of the project that pose the most risk when they are forgotten.

Architects who recount their greatest career disasters often tell stories like the following:

- Completely scaling back a project for no additional fee because the original plan was far beyond the owner's budget
- Discovering late in the construction documents that the proposed addition extends into an easement
- Hearing the owner comment at any point in the project, "I don't know why this is in the plan—I don't want this."

Architectural practice disasters like these occur for all kinds of reasons, including the changing whims of mercurial owners. Architects can do much, however, to protect themselves from harm by devoting early attention to the project, such as listing the fundamental governance issues of the project and gaining the owner's agreement regarding them. This is often referred to in many endeavors as the process of persuading the owner to *buy-in* to the program.

THE GOALS

Architects often prepare project programs for owners: narrative descriptions of the spaces required, their relationship, and attributes of the facility. In effect, the program is a statement of goals. Not every project requires a formal written program, but every project of any size—complex or basic—deserves a well-defined set of goals. Start by listing five fundamental goals for the project. This list defines the owner's highest, most important priorities. He likely has others, but these are the goals the architect must deal with above all others. Here's a basic example for a small private school addition:

- New classrooms: Add two new classroom spaces, each 800 sq ft.
- HVAC: Separate thru-wall systems for each classroom.
- Technology: Electronic marker board in one classroom, and flat-screen monitors in both.
- Plumbing: Install a lavatory and countertop in each classroom.
- Environmental: Use energy-efficient lighting and green products.

The list seems simple, but it is a roadmap for the architect and owner to follow in reminding themselves of the main goals they must meet as they move deeper through the project. The architect should discuss this list with his client, and then prioritize it:

1. Add new classrooms.
2. Add new HVAC units.
3. Add new lavatory and countertops.
4. Use energy-efficient and green products.
5. Add technology components.

Even in this basic example, the architect can find some guidance to aid him in setting up his bid documents. Because he placed environmental concerns ahead of technology, the client is telling the architect that he considers this aspect of the project more important. If he is facing a very tight budget, the architect may respond to this information by including the electrical rough-in for the technology components in the base bid, while making the flat-screen monitors and electronic marker board add alternates. Since even the environmental piece is a lower priority than the basic needs of the space (area, HVAC, and plumbing), and green products sometimes cost more than less environmentally-friendly products, the cost-conscious architect may also create separate specifications and add alternate bids for environmentally-sensitive products versus those that are not. Once again, if money is tight, the owner will have the option of deciding whether to upgrade to green products or abandon them in favor of higher priorities.

After defining the goals of the project, it is essential to review the resources available to meet those goals. The most important among these is the budget. Occasionally, owners of small projects are not very candid in discussing their budget expectations with the architect. Whether out of fear that the architect and contractor will collude to cost him more than it otherwise would, or out of a belief that project costs will expand to meet their budget, there is a clear reluctance among some owners to discuss costs early in the project. This is dangerous to the architect and to the success of the project. The budget is the main driver of decision-making on any project, and an owner who is not comfortable discussing money with his architect is jeopardizing his project. When he senses this reluctance on the part of the owner, the architect must force the issue by creating a draft budget for the owner of likely construction costs for the size and type of facility being proposed. After explaining that these are budget figures for similar facilities in his area, the architect should ask the owner to confirm that a budget of the approximate amount in the example is acceptable to him. If the architect does not feel confident enough in preparing an early budget estimate for the owner, he should provide him with a small range of costs per square foot pulled from a standard cost-estimating reference. At the very least, the architect needs to be assured in some way that the owner's budget expectations for the project are not unreasonably low. Discovering this fact during construction documents—or worse, when the bids are opened—is an unnecessary risk and may occur too late for a cure.

Architects sometimes argue that not having a defined budget from the owner protects them from having to redraw an over-budget set of plans at their cost. This argument of "I can't be responsible for what I don't know," does not usually wash with owners. They always have a budget, will usually consider it reasonable despite evidence to the contrary, and will always expect the architect to design within their budget, whether or not they shared it with him. Delaying the day of reckoning until deep into the project by not having a frank discussion with the owner regarding his budget rarely results in a happy outcome. For every owner who will agree to pay the architect to revise the plans to meet his newly-revealed budget, there are many more who will insist that the architect should have made him aware of the *true* costs earlier, and should have designed in the least costly manner to begin with. Even if an owner does not include project budget preparation in the scope of services for the architect, he should include it himself—as a protection against the unknown.

A second basic resource is time. Even sophisticated owners can be surprised at the amount of time required to obtain governmental approvals, and for design, detailing, bidding, and construction of a facility. Architects benefit from helping the owner to understand early in the process how much time the normal cycle requires, including an estimate of construction time. As with the budget, owners can have unreasonable expectations about how quickly they can have a new building under construction. In particular, clients can be unsympathetic to the time required by architects and engineers to prepare the construction documents. Once the design is finalized, they often fail to understand the amount of work—and time—required to create a set of quality documents. For the architect, therefore, it is important to engage in some salesmanship to help the client understand the value of spending time on the documents, and to determine early if the likely schedule—from program to punchlist—is acceptable to the owner.

The best way of accomplishing this is by creating a preliminary project schedule. A project schedule is distinguished from a construction schedule in that it encompasses the entire time required to create the building from the owner's perspective—from design through closeout. This schedule should include an estimate of time required for every component of the project. Seeing the time required, and the relationships, for construction documents, permitting, bidding, and construction helps the owner with unreasonable schedule aspirations to understand why construction takes as long as it does. For the architectural portion of the preliminary schedule, the architect should stress the importance of time spent on the construction documents by including milestones for review periods at the end of design development, and again at 50 percent and 90 percent of the construction documents. The project schedule should include time for owner review, highlighting the importance of spending time to make sure the documents meet the owner's needs. See Figure 2-1 in Chapter 2 for an example of a project schedule.

Along with ensuring that he and the client are in agreement on the project budget and schedule, the architect should also devote time to making sure he understands the regulatory or legal restrictions for the site, and other physical site constraints that affect his work. If a civil engineer is engaged by the owner, the architect should review and understand the setbacks, easements, runoff management basins and structures, utility locations, and soil conditions present on the site. If a geotechnical report is advisable, the architect should inform the owner and help him obtain these services. If no civil engineer is required on a small project, the architect should ask the owner for a thorough topographical, utility, and boundary survey, showing all the information available on the site. Where a civil engineer is involved, the architect should visit the site with the civil engineer's plan in hand to see how the grade moves along the proposed building line. Building entry steps and a ramp along a steeply sloping drive may gain zoning approval but result in an architectural and functional headache. Working with the civil engineer early in the project can save frustration and difficulty for the architect down the line.

As a final step in preparing to design, the architect should perform a quick code review. Why so early, before even a line has been drawn? Because project peculiarities can lead to code oddities that affect the cost and schedule of a project. Some real-life scenarios:

- The plumbing code for a fitness facility requires an extraordinary number of shower facilities, necessitating a variance appeal to the building department;

- Adding a large addition to a retail store exceeds the maximum building area, requiring the creation of a new fire area and building separation;

- Rear exits out of a movie theatre occur below grade, creating a problem for barrier-free egress.

None of these problems are insolvable. They represent the types of problems that architects wrestle with all the time, on a wide range of projects. It is advantageous, however, to know about them well ahead of time. The architect of the fitness facility will have time to document and apply for a variance to reduce the number of shower stalls. The architect for the movie theatre can work with the civil engineer to improve the rear exit grade situation on the site plans and avoid barrier-free code compliance problems.

TIGHTENING THE CONSTRUCTION DOCUMENTS

Architects do not always utilize the best tool at their disposal for minimizing their risk—their documents.

- *Cover the scope of work:* Make sure all work scope is clearly covered; delineate between new and existing.

- *General to specific:* Set up your documents to lead the contractor through the work in a logical manner. Use key plans, building structural grids, and other devices to orient the bidders.

- *Key down:* Key clearly from building sections and plans to smaller details.

- *Coordinate:* Take time to coordinate among architectural and consultant drawings.

- *Extra paper:* Do not try to communicate too much on one page; copy a plan, elevation, or section to show a new layer of information.

- *Fresh eyes:* Have someone unfamiliar with the project spend a few hours looking over the drawings and specifications for gaps and clerical errors.

- ***Peer review:*** Depending on the nature and complexity of the project, it may be helpful to obtain a review by an outside consultant or other party for quality control purposes. Where the owner has engaged a construction manager, their scope may include a constructability review of the documents.

There are a number of common construction document errors that tend to result in change orders:

- Incorrect scale on drawings (a common problem with less conventional $^3/_{16}$ or $^3/_{32}$ inch scales)
- No detail of custom conditions
- Extent of finishes is unclear (particularly on interior elevations)
- Lack of coordination among architectural, structural, mechanical, plumbing, and/or electrical drawings
- No plumbing runs shown for plumbing indicated on the architectural plan (roof drains and drinking fountains are common omissions)
- Structural detailing necessary for architectural features or roof-mounted equipment
- Mounting heights not clearly indicated
- Hidden conditions not considered in bid documents
- Standard details do not reflect actual conditions or work shown elsewhere in the documents (door and window details are a common example)
- Interface of new and existing construction not detailed
- Separation of new versus existing to remain construction not clear
- Architectural issues relating to grading are missing exterior steps, ramps, and pads, and exposed foundation walls
- Water management issues, such as missing scuppers, overflow drains, gutters, downspouts, and underground connections
- Power irregularities (mismatched phases or voltages on civil, electrical, and HVAC drawings)
- North arrow delineation.

A thorough review of the construction documents is tedious work, no doubt. But using a fresh pair of eyes inside, or outside, the architect's office can be invaluable in catching those errors that can be so embarrassing and costly once construction begins. Such reviews are often cited in court as supporting the diligence of the architect in performing his work consistent with the standards of professional care.

SPECIFICATION PROTECTION

Specifications are often considered by architects to be the construction document's dental drill to the drawing's sledge hammer. The drawings do all the heavy scope lifting, with the specifications filling in the product and quality details. In the risk management area, however, the specifications are the workhorses in protecting the architect and owner from harm. To use them to their full advantage, the architect must understand the benefits of both the specific and general nature of various specification sections.

GENERAL SPECIFICATION SECTIONS

Sometimes the least specific sections of the specifications are the ones that provide the most benefit to the architect. Following is a list of specification sections that are critical risk reduction tools for any set of construction documents:

- *General conditions:* The most efficient means for the architect to cover general conditions requirements is by referencing Document A201©, prepared by the American Institute of Architects. This document is a comprehensive, case-law tested set of requirements that is used throughout the industry and is usually well understood by architects, engineers, and contractors alike. Attorneys will argue that aspects of the AIA General Conditions do not adequately protect the owner and architect in specific situations, and the architect should seek guidance from the owner and his attorney in modifying sections that they consider onerous. Because A201© is a general document, every section does not work for every project. Architects should include in their specifications a section modifying the general conditions for the needs of their particular project. This is usually accomplished in the form of amendments to the general conditions.

- Supplementary conditions: This section consists of additions to the general conditions, which often include: laying out the specific work conditions for the project, including the hours of operation; noise, dust, and vibration restrictions; special safety concerns; and any other requirement specific to the project.

- Cutting and patching: A useful section for both new construction and renovation, this section lays out appropriate methods of cutting and patching openings for systems runs, and puts the contractor team on notice that clean procedures are a requirement of the project.

- Reference standards: This section establishes a basis of quality in manufacture and installation for a wide range of products. This is particularly important in finish or aesthetic installations, such as tile installation, and precast or poured-in-place concrete, where proper procedures are vital to the quality of the finished product.

THE FRONT-END

The front-end specifications are often referred to as the "boiler plate," or Division 0 specifications. These sections of the specifications define the basic relationships, rights, and responsibilities of the various parties to the agreement (typically the owner, architect, and contractor), as well as how the agreement will be administered. Public and corporate owners who contract for a large volume of construction work and capital projects often develop their own set of front-end specifications, tailored around the needs of their projects and the way they prefer to work with design professionals and contractors. Others rely on the architect to customize standard front-end specifications to their individual needs.

Front-end specifications set the rules of the road for the contractor. Collectively, they represent a list of operating rules and procedures the contractor must factor into his proposal and follow during the course of construction. Following is an expanded list of front-end specification sections. Some of these sections may not be necessary on typical privately-bid projects, but the list provides a sense of the expansiveness of the boilerplate section of the specifications:

- Advertisement for bids
- Instructions to bidders
- Bid proposal form
- Certificate of ownership
- Bidders affidavit
- Qualification questionnaire
- Non-collusion affidavit
- Site visit form
- Form of bid bond
- Consent of surety
- Affirmative action affidavit and plan
- Affirmative action plan
- General conditions
- Supplementary general conditions
- Contract form
- Performance and payment bond
- Form of guarantee.

Of the front-end specification sections listed above, two are particularly important, and are the most scrutinized by bidders:

General Conditions

Generically, the general conditions of construction (commonly shortened to "general conditions") lay out the rights, responsibilities, and relationships of the parties involved in the construction of a project. If this sounds like a monumental task, that is because it is. The manner in which owner, architect, and contractor deal with each other during the course of construction has evolved (some would say devolved) over many years of practice and thousands of instances of case law into an accepted program of industry practice. For each architect to document his own set of general conditions would be time-consuming and inefficient. For this reason, most architects use those provided by the American Institute of Architects (AIA)—specifically the AIA's document A201©. But because various situations in each relationship and on each project differ, architects often include sections in the front-end specifications that modify the AIA general conditions for their particular needs. Some of the areas commonly adjusted include:

- Whether the contractor or owner pays for permits
- Provisions regarding insurance and bonds
- How changes in the scope of work (change orders) are approved
- Owner project representation and responsibilities of the architect.

Supplementary Conditions

The second important section of the front-end specifications that is of particular interest to contractors is what is called the supplementary conditions. This section supplements the general conditions (hence the name) by zeroing in on specific requirements surrounding the project, people, and scope of the owner's work. Typical supplementary condition items include:

- Time of completion
- Coordination with owner's own forces
- Guarantee or warranty provisions
- Temporary utilities or temporary heat
- Cold or hot weather procedures
- Safety and public protection
- Field office requirements

- Site access and control requirements
- Site signage requirements
- Project meeting and documentation requirements
- Site cleanliness, recycling, or disposal requirements
- Geotechnical data or soil management requirements
- Project management documentation requirements
- As-built drawing requirements.

BID STRATEGY

The process of securing the firms responsible for constructing the project is commonly referred to as bidding. Given the many alternative means of obtaining a contractor used by owners, a better word might be "procurement," but bidding remains the most common method and the one most prone to causing liability for the design professional. The use of the word *strategy* in conjunction with bidding indicates there is a pre-conceived scheme, an organized plan to structure the bidding in such a way as to ensure that the owner obtains responsible bids that are within his budget. This should be the case with any set of documents that it being bid to contractors, but this is not usually the case. If it somehow seems unsavory to be planning the bidding to this extent, consider that contractors strategize heavily in preparing their bid responses. They include an assessment of their competition, construction market conditions, the owner's likely bid alternate preferences, and the potential for change orders in preparing their bid response.

DRAWING NOTE PROTECTION

Drawing notes can be a useful tool in alerting the contractor to special conditions or requirements in the construction documents. They are most effective when they are specific and singular. Notes should apply to a particular instance of concern and offer the contractor specific guidance or instruction in that area alone. General notes that are peppered throughout the drawings with reference to a specific area, and which require the contractor to perform work beyond his normal scope, are generally ineffective and treated as little more than wallpaper. Here are some examples of overused general notes in construction documents:

- Field verify that all drawings conform with existing conditions
- Contractor responsible for ensuring adequacy of all framing
- Ensure that all construction is in accordance with all applicable codes
- Contractor to perform work with no disruption of existing operations.

Architects should avoid using drawing notes that include words such as *all, confirm, certify, ensure, guarantee, or complete.* Comprehensive notes do not obligate the contractor to perform work past the normal standard of care required of him in the general conditions, and because they are so broad, they do not alert the contractor to those specific conditions where special care on his part is indeed required. Where those situations occur, the architect is much better served by inserting a note stating exactly his concern and how the contractor can address it. If an architect, for example, is concerned with a particular area of an existing facility where he had a difficult time obtaining accurate existing conditions measurements and he wants the contractor to verify the dimensions before he begins framing, he should simply state this fact: *Field verify existing dimensions between column lines B and C. Dimensions shown on the drawing may require adjustment to conform to existing conditions. If so, add or delete max. of 6 inches from the center office. Contact architect if discrepancy exceeds 6 inches.* In this example, the architect provides a specific note that applies to a specific situation. He tells the contractor what to check in a limited area, how to respond to what he finds in his framing layout (within parameters set by the architect), and alerts him to contact him if the discrepancy is more than anticipated. This is a reasonable, legally defensible request of the contractor that, due to its specific nature, he is likely to take notice of and act on. A general note telling him to "field verify all dimensions" accomplishes little in alerting the contractor to the architect's real concerns and how to resolve them. This example applied to a particular area of concern in an existing building. What if the architect is concerned with a general firewall situation where it is critical that all penetrations be sealed in an appropriate manner? A general note addressing this type of situation might read: *Contractor responsible for ensuring all firewall penetrations are sealed in an acceptable manner.*A more specific note, however, clues the contractor into the architect's main concern: *Penetrations in the two-hour firewall must be sealed according to details on Sheet A8.1 and in Spec. Section 07800. Fire putty alone is only acceptable on openings smaller than 2 inches.* A contractor reading this note understands fully that the architect has probably been burned before by a contractor using fire putty alone to seal openings that require other measures. He now knows the concern, the architect's heightened awareness of the potential problem, and what he is specifically being asked to do to avoid it.

COVERING PROJECT SCOPE

Nothing is more critical in producing complete construction documents than covering the full project scope. Nothing is more embarrassing—and potentially costly—to an architect than having to explain to a client why part of the project did not appear clearly in his documents. It may sound incredible that an architect who has spent months developing a set of construction documents could leave out a portion of the work. It is actually an easy mistake to make (see Risk Hazard Flags for a list of common scope omissions). Consider that someone working intensely on a project, drawing details, and writing pages of specifications day after day, can easily overlook a significant part of the project they have always known to be included and assumed, as a result, that they had clearly represented it in their documents.

Scope problems occur in two areas: 1) *The bulk problem:* Is the scope missing? and, 2) *The extent problem:* Is the extent of the scope clear? Bulk problems occur when architects forget that they are presenting the drawings to bidders who are seeing them for the first time, and making quick judgments as to what is and is not part of their responsibility. These bidders have not attended all the meetings, have not heard the discussions, and know only what they see on the documents. When an architect assumes that

Risk Hazard Flags

Common Scope Omissions in Construction Documents
- Hidden conditions
- Closures or intersections
- Finish extent ambiguity
- Exterior elevation areas not shown
- Repair and renovation work scope
- Difficult details that never get drawn

a section of an existing roof is to be replaced but does not clearly indicate it, the roofing contractor misses it on his quantity takeoff. When an architect intends for a brick veneer to be applied on a portion of the building not shown on an elevation, the contractor assumes it is painted masonry because he has no indication to the contrary. These two examples of missed scope stem from an architect's extreme familiarity with a project and not translating this knowledge into clear instructions on the documents. Assumptions and knowledge built up over a long period convinced the architect that the information was as clear to contractors as it was to him. Fortunately, the solution to this problem is a simple one. Having a set of *fresh eyes*—someone not involved in the details of the project—review the documents before they are issued can catch many of these omissions. To be effective, the reviewer must be experienced in construction documents (not a job for the summer intern), and specifically look for scope clarity. Following are some common areas where project scope is often missed, or communicated poorly, in construction documents.

Hidden Conditions

Refer to Chapter 7 for information on how to deal with hidden conditions. See the C.A. Anecdote for an example of how an architect can benefit from anticipating scope even in hidden conditions.

Difficult Details

Some details are simply difficult to figure out. This is particularly true of closure and intersection details, where it can be difficult to visualize or even know for certain how the building elements will come together. Some architects tend to shy away from detailing what they are not sure about, feeling perhaps that to detail something that is incorrect is a type of architectural fraud. Contractors and owners do not view it that way, however. They want to see situations covered in some manner. The contractor will be quick to point out that "he can't price what he can't see," and the owner will be unreceptive to paying a change order for a recognizable condition that, in his view, the architect simply chose not to deal with.

The solution? Draw something. The architect must establish some value in the construction contract for work that must be done. If a particular detail simply cannot be figured out, he must do his best at covering the scope in either detail or note form. Either way, the contractor must be made aware that this scope exists, the relative extent, and how he must deal with it. In defense of full disclosure, the architect can note on the detail: *Final conditions may differ from detail. Contractor to assume this scope in base bid and contact architect if conditions differ.*

The contractor will be quick to contact the architect when the actual conditions do not line up with the detail. At that point, the architect and contractor together can work out an appropriate detail. As a result of the *placeholder* detail, however, there is value in the contract to apply toward the real solution. Even if a change order is required, it will be smaller than it otherwise would have been. Of course, the owner must be aware of this strategy to cover an unknown situation. Additionally, work priced during the bid period is generally considered to be lower priced than that priced as a change order afterwards. In fact, statistical studies have shown that additional work priced during the course of construction as a change order is almost always higher than the same work priced during the bid period.

Coordination Problems

Coordination among his consultants bedevils many architects. Most, at one time or another, have faced change orders resulting from a

C.A. Anecdote

The Forgotten Canopies

The Problem

When the bi-weekly job meeting was over, the contractor's field superintendent and project manager asked Tom to look at a situation with them.

"We were concerned from the start about what we'd find when we removed the canopy sheathing," said the field super. "Take a look."

Tom shielded his eyes from the sun and looked at the framing in an open canopy. It was a true mess, a conglomeration of different sized framing members forming a strange and odd-looking truss. The contractor's carpenters had exposed three other canopies (out of a total of dozens), and each one was framed just as haphazardly.

"This is a mess," said Tom. "It looks as if they gathered up all the scrap lumber from the rest of the project, gave it to an apprentice carpenter, and said 'build some canopies.'"

"We're thinking every one of these should be demolished and re-framed," said the project manager. "If the rest are this bad, we're looking at a substantial amount of extra work."

"And a substantial change order as well, I assume?" said Tom.

"It's a hidden condition. The owner will understand no one could have predicted something this bad."

"You don't know clients like I do," said Tom. "No owner likes change orders, no matter how defensible we think they are. Besides, this won't be a total change order."

"How do you figure?" said the project manager.

"The construction documents show a section with new canopy truss framing, and the bid documents included an estimated quantity of ten new canopies framed with that detail."

"I saw that detail," said the field superintendent, "but I assumed it was meant to reflect assumed existing conditions."

"Nope," said Tom. "A note on that sheet tells you it is new canopy framing to be used in the event the existing was deemed inadequate. We guessed we would find a certain number of conditions with deteriorated framing due to water damage from poor flashing. Of course, no one thought we would find anything this bad or this extensive."

"I didn't see the estimated quantity in the bid documents," said the project manager. "We didn't include it in our bid."

"I'm sorry you didn't, but it was clearly referenced in the bid proposal form, and the Contract for Construction included all bid documents."

The Resolution

The project manager reviewed the documents and agreed that his company owed the owner ten reframed canopies within their agreement. Their change order value for reframing the balance of the complex's canopies was much higher than the Tom anticipated. It appeared evident that the contractor was trying to recoup some of the money lost on the unexpected framing through the change order. Tom regretted not including a unit price bid for the canopy reframing as part of the bid documents. A unit price would have provided the owner with a predictable value for the work, and perhaps have helped to clue the contractor into the existence of the estimated quantity and the importance of the framing detail. Tom negotiated with the contractor to obtain a more favorable change order for the owner, but still felt like he could have anticipated this hidden condition better.

failure to coordinate among all his various engineering consultants, with the result that an item shown on the architectural drawings is not picked up elsewhere in the documents: Roof drains on the architectural roof plan that have no plumbing on the plumbing plan; Light fixtures on the architectural reflected ceiling plan that never appear on the electrical lighting plan; Rooftop units with no structural dunnage or framing to support them. Architects argue that the contractor is responsible for the entire set of documents, and therefore should ensure that the plumber prices out the runs for the roof drains. The plumbing sub will respond that he only looks at plumbing and expects to see all his scope on those documents. The contractor will argue that he and his sub cannot bid piping that is not defined. At best there is a time-consuming argument and embarrassment for the architect. At worst, the owner ends up having to pay for plumbing to roof drains that he feels (with justification) should have been in the job from the beginning.

The solution is coordination. Architects must take time to talk with their consultants, ensure they are working off the latest base plan, and review their documents (however briefly) when they receive them.

Extent of Work

A common problem is a failure to show the full extent of work in the documents. This occurs mostly in the area of exterior and interior finishes. Architects do not usually create elevations for every surface of a building. Rear additions, roof structures, odd angles, and surfaces screened by others in an elevation can end up not being depicted anywhere clearly in the documents. The architect must be cognizant of this situation and, if he does not have the time to draw the elevations of the areas in question, he must at least convey the finish information with a plan or elevation note.

Other finish extent problems in the documents can occur with floor finishes, particularly in transitions and elevations. Mud-set tile or wood floors on sleepers may meet other finishes at significantly different heights. The architect must detail the transition between these finishes, and if the transition requires special barrier-free provisions, detail those as well. More commonly, architects rely on a finish schedule that simply does not clearly indicate where finishes stop and start, particularly at transition areas such as interior ramps and stairs. Where possible, a finish floor plan is advisable in any project where a number of different floor finishes are being utilized.

Repair and Renovation Work

Documenting miscellaneous repair and renovation work is difficult. Architects can generally recognize that some element of existing construction must be reframed, reclad, jacked up and supported, or otherwise improved. It is not always easy to capture this scope in the construction documents. Part of the dilemma is a natural wariness of making the problem larger than it really is. What may start out as a simple carpentry problem to the contractor can morph into an expensive structural repair once the architect and structural engineer finish documenting it. Another worry in overly describing how much work to do in repairing a piece of construction is the danger of never saying enough. Renovation work, by its very nature, has an ability to expand beyond all reason. Take down one thing and the contractor will inevitably discover two other items that require repair. Items requiring repair generally must be repaired. But owners of renovated facilities that have been neglected over a long period usually must learn to live with the fact that they cannot afford to have everything repaired as new. This inevitably requires a certain amount of creativity and cooperation on the part of the contractor, and a willingness on the part of the owner to recognize that he must pay some extra amount to repair deteriorated work that was ignored for too long.

For the architect, documentation of repair work is probably best handled counter to everything else you will read in this book. Less is more. The simple *repair as required* note has served architects well for a long time, and its nebulous quality is probably the best argument for why it is effective in renovation work. However, if a specific level of repair is expected, the architect must specify it rather than leave it to chance or the contractor's expectation that he must simply match the surrounding quality.

Intersections and Closures

It is easy to forget those odd conditions on a project—new construction or renovation—that deal with the awkward areas where building components intersect with one another or require closure. Unusual roof offsets and extensions, peculiar wall intersections between dissimilar surfaces, a difficult meeting of angle surfaces all require some detailing on the drawing but are often overlooked. These are not usually a significant source of change orders on a project, but they nevertheless take up an architect's time during construction and usually require a quick decision. For that reason alone, it is worth spending time during the preparation of construction documents to identify as many of these conditions as possible to avoid having to deal with them piecemeal under the rushed conditions of construction.

Risk Reduction Tools

Construction Documents Checklist

- Accurate table of contents
- Drawing issue dates all match
- Property location clearly noted
- Owner and professional team information shown
- Building code compliance information, including egress, occupancy, and fire ratings shown
- Plumbing code compliance information shown
- Plan scales consistent and accurate
- Exterior stairs and ramps coordinated with civil
- Exterior elevations of all walls included, with approximate grade shown
- Exterior elevations include finish type and extent
- Roof plan shown, including drainage provisions
- Roof edge flashing, penetration, and parapet details included
- Reflected ceiling plan included
- Egress lighting, exit signs and smoke/heat detectors shown
- Finish schedule and plan included
- Window and door schedules included
- Hardware schedule or specifications included
- Partition schedule shown and keyed to plan
- Sufficient dimensioning shown on plan
- Building sections and wall sections keyed correctly on floor plan
- Wall sections shown for each different type of condition
- Wall sections keyed correctly on exterior elevations
- Necessary access panels, roof hatches, and venting shown
- Structural sections align with wall sections
- Details shown for necessary building elements
- Details keyed correctly on building and wall sections
- Structural documents include provisions for roof top equipment
- Cold-formed framing clearly indicated or made responsibility of contractor engineer

- MEP documents coordinated with architectural and structural disciplines
- Gas and electrical load assumptions included
- Plumbing fixture schedule included
- HVAC equipment schedules included
- Plumbing, gas, hot, and cold water riser diagrams included
- Light fixture schedule included
- Switching and circuiting shown
- Electrical panelboard schedule included
- Fire protection scope shown
- Specifications include front-end sections
- Specifications include quality reference standards
- Specifications include all products shown in drawings
- MEP specifications included
- Structural specifications created or checked by engineer
- Required submittals and tests indicated in specification sections

CONSTRUCTION DOCUMENT CHECKING

Checking the construction documents is painful duty. It is difficult for an architect to sit down calmly with a set of documents that he has been working with for a long period of time and go over them sheet-by-sheet, detail-by-detail as if for the first time. Considering that this check must be performed sufficiently in advance of the drawing issue date to allow correction of any errors that are discovered, one can get a sense of why construction document checking has become a lost art in many architectural offices. (See the Risk Reduction Tools for a construction documents checklist.)

The hard reality is that the small amount of time spent checking construction documents results in the biggest payback an architect can expect on a project. Even the most inexperienced, fresh-out-of-school intern in an architectural office can spot many of the fundamental errors that can bedevil an architect during construction. Within two or three hours, a fresh set of eyes can pore over a set of documents to check for mis-keyed details, out-of-date consultant bases, illegible or missing schedule information, and title block irregularities. The time spent responding to even the simplest of field complaints—a missing detail for instance—can make the in-office time spent in reviewing the documents worth the effort. There are much better reasons to conduct an in-house review, of course, and those reasons center on avoiding professional liability. Consider the following example.

An architect is called by the project manager.

"We have a problem," he says. "The plumbing subcontractor bid 25 percent less gas piping than is required for the project. He wants an $18,000 change order."

"That sounds like his problem," said the architect.

"He says he bid using the $^1/_8$ inch=1 foot scale shown on the plumbing roof plan. He later discovered that all the plans in the set are actually $^3/_{32}$ inch=1 foot.

"He's supposed to review the entire set of documents, which would have alerted him to the error, and not just slap a scale on the drawings."

"I agree, to an extent, but he's a plumber. Plumbing drawings are all he looks at. Besides, why didn't your engineer show the correct scale? Then we wouldn't be having this discussion."

In this scenario, the architect will now struggle to mitigate a large change order situation resulting from a very simple error: a consultant failing to show the correct scale on a plan. The fact that this error could have been caught by even the most rudimentary of drawing checks indicates the importance to the design professional in taking the time to review the construction documents before he puts them out on the street.

Following is a list of basic and quick items to look for in the construction document check:

- Do the consultant bases match the architectural base?
- Does the cover sheet show all sheets in the set?
- Do all title blocks reflect the correct project information, same issue date, correct scale?

- Are all wall sections keyed correctly to the plans?
- Are all details keyed correctly to the plans and wall sections?

CREDITS IN THE CONSTRUCTION DOCUMENTS

Change orders are a normal part of the construction process, resulting from the fact that no set of drawings can completely and fully document all the work required to construct a building of any complexity. Any number of decisions, interfaces, and components, as well as the human factor of interpreting the bid documents differently from how they were intended, inevitably results in claims for more money. Owners can be more or less understanding of change orders, depending on their experience and temperament. When a responsible contractor makes change order claims that are within reason and fairly valued, architects are normally able to explain the situation to their clients and work out a solution. When contractors or their subs submit change orders that the architect and owner consider petty or abusive, when a flood of change orders arrives at the end of the project, or when a contractor submits a large change order for additional general conditions as a result of delays caused by the owner or his professional team, attitudes harden quickly. Change orders of these types tend to be aimed mostly at the architect and his consultants for *errors and omissions* in the construction documents. They allege that work was not shown (omission), was shown incorrectly or incompletely (error), or that the architect failed to respond to requests for information or submittals quickly enough to allow the contractor to maintain his schedule. When such change orders begin rolling in, the architect will have to devote considerable time to sorting through the claims for extras and working to defend himself and his client.

Money in the Bank

At those times, it's nice to have money in the bank. It is possible to build into the construction documents, buried deep in the specifications, requirements that are likely to be either missed or ignored by the contractor and his subs during bidding, and will never be carried out during construction. This means, technically, that a requirement of the contract was not performed and the owner is entitled to a credit for the value of the unperformed work. The requirements that lend themselves to this technique are procedural, related to the field administration of the contractor or documentation associated with his monthly application for payment. They provide little value to the owner, are tedious to comply with, and their elimination by the contractor does no harm to the project or life safety in any way. Still, they are requirements, and the architect can pull them out when he senses the need to restore some balance to the change order situation.

Abusive? Using credits of this nature to punish an otherwise well-performing contractor or to falsely burnish an architect's change order percentage on a contract would be abusive. One could also argue that using surreptitious credits in this manner to defend the owner against suspect change orders, while a worthy goal, is countering one wrong with another. In a world of absolutes, this would be true. But when owners are facing large change orders aimed mostly at the architect, it is advantageous to pull from the drawer a handful of credits to restore a rough sense of equity. The effect can be remarkable. A contractor who previously held a large claim, and a reasonable expectation that at least some of it might be paid, suddenly finds himself subject to counterclaims appearing straight out of the documents he is using to make his argument. These credit claims accomplish another important goal. Appearing almost out of thin air, they remind the contractor that the architect was the author and reigning expert on the construction documents. A shrewd architect might hint to the contractor that where these came from others may follow.

By itself, each credit does not constitute a large value. In fact, the value of each one can be very speculative, dependent on interpretation. Their effectiveness comes from the fact that they are mostly related to continuing expenses of field operations, which means a relatively small value is multiplied week after week, month after month, for the entire course of construction. That, plus the fact that the architect should include three or four of these types of credits in the specifications, will add up to a sizable sum at the end of several months of construction.

The goal in using these credits is not to unjustly enrich the owner, but to persuade a contractor who is submitting claims of dubious value that he should moderate his requests and settle equitably. The credit claims are simply a tool for the owner and architect to use in promoting something that should happen anyway.

Examples of Hidden Specification Credits

- Larger than normal field trailer, with dedicated office and telephone line for architect and owner
- Bi-weekly photographic documentation of progress; min. 30 photographs; reproduce two copies of $8\,^1/_2 \times 11$ color images per submission
- Construction site to be enclosed at all times during the course of construction with an 8 ft high chain-link fence
- Contractor to maintain adequate field supervision (defined as more than one superintendent) on site during all working hours
- Contractor responsible for providing dedicated fire watch (not subcontractor personnel) in any location where open flames or temporary heat is used
- Temporary lighting requirement
- Custom project sign requirement
- Limited working hours in the specifications extended by the owner (saves the contractor general conditions costs).

CONSTRUCTABILITY REVIEWS

Constructability reviews consist of either an independent or in-house review of construction documents at a high level of completion (90 percent or greater). The most effective reviews are carried out by people who are knowledgeable in construction means and methods and who possess some architectural and construction experience, but were not directly involved in preparation of the documents they are reviewing.

Anyone who has worked closely on any project over a period of time realizes that there is great value in laying it aside for a cooling-down period, getting away from it, and taking a fresh look at it later. There is not often time to do this in a busy architectural office under the pressure of a bid deadline. There is always an opportunity, however, to have another experienced architectural or construction professional in the office (or out of it) look over the documents for those obvious flaws that only a fresh pair of eyes can spot. Flaws such as the missing flashing details, the roof drains with no plumbing, the footing detail that does not extend below grade, the rated wall that isn't, or any one of hundreds of other errors that turn up on working drawings. The constructability review is not the same as construction document checking. Routine checking of the set should be performed by an in-house person intimately familiar with the documents. Ideally, this person will catch all the "easy ones:" the mistakenly keyed sections, the incorrect scales, mislabeled keynotes, and the like. This type of review is very helpful, but it is not done for the purpose of improving the design intent. The constructability review is something entirely different. It is a high-level review by a person unfamiliar with the documents, looking at them for the first time as a bidding subcontractor or general contractor would do. The constructability review is a quick one, scanning for the big misses, the scope gaps, the work that needs to be priced but is not clearly shown. The obvious intent is to catch those missing elements that would be caught anyway as dozens of bidders pore over the documents.

The less apparent benefit, however, is that the constructability review may catch some of the most painful problems in the pantheon of architectural errors and omissions—coordination problems. One of the least savory secrets of the architectural profession is the scant amount of time the architect spends coordinating his drawings with those of his consultants. Except it is not a secret at all. Most contractors and subcontractors have a wealth of stories about variations, large and small, among the consultant drawings in a set of construction documents. The consultant drawings come rolling in via e-mail just before the final deadline for the project and the architect has little or no time to review them. He sent the consultants the updated base plan and copied them on all the e-mails. They had the information, so the drawings should be right, he reasons.

The constructability review should be performed on a printed set. Why? Because a test print will flush out readability, line weight, title block, and layer issues that can be caught before they turn up in the bid sets, but which are not easily discernable in electronic form.

Following is a basic list of what to look for in the constructability review.

1. **Architectural:**
   a) Completed site plan
   b) Completed floor plans, elevations, and sections
   c) Architectural wall sections and details completed and keyed
   d) Finish, door, window, and hardware schedules completed, including all details
   e) Reflected ceiling plans completed
   f) Code compliance information clearly shown
   g) Wall (partition) types clearly indicated, including location of fire separation assemblies.

2. **Structural:**
   a) Structural foundation, framing plans completed
   b) Structural sections and details completed
   c) Structural calculations completed.

3. **Mechanical:**
   a) Large-scale mechanical details completed
   b) Mechanical equipment schedules completed
   c) Complete energy conservation/usage calculations included.

4. **Electrical:**
   a) Lighting and power plan, showing all switching and controls
   b) Light fixture and electrical equipment schedules included
   c) Distribution information on all power-consuming equipment, including lighting, signal, communication device(s), and branch wiring completed

    d) Tele-data locations and equipment information included

    e) Life safety components indicated and specified, including emergency lights, exit signs, and horn-strobe devices

    f) Electrical load calculations completed.

5. *Civil:*

    a) All site plans, including parking and roadway information indicated

    b) Site electrical, storm and sanitary sewers, water, and other utilities shown.

6. *Landscaping:*

    a) All landscape, hardscape, exterior lighting, and irrigation plans complete

    b) Site amenities shown and specified.

COST ESTIMATE PREPARATION

Architects are often required to provide cost estimates at one or more stages of the project. Though sometimes viewed by architects as a combination of voodoo and Ouija board management, the preparation of a cost estimate can be a useful exercise in understanding where the value lies in the construction, and conversely, where potential savings can be realized if necessary.

The title of an estimate should reflect the level of detail associated with it. For instance, if the architect is preparing an estimate based on a schematic or preliminary set of plans, and outline specifications, the title should reflect the basis of the estimate:

- Preliminary Construction Budget Estimate
- Schematic Construction Cost Estimate
- Preliminary Estimate of Probable Costs

An estimate based on complete construction documents might be titled:

- Final Statement of Probable Construction Costs
- Final Construction Cost Estimate

In all cases, the estimate should be dated and clearly state the documents used as the basis for its preparation. An estimate should include a disclaimer, noting this is an estimate based on the time and market conditions when it was prepared. Changes in either may result in increased costs.

Sample Disclaimer

This document represents our opinion of the probable cost of the construction based on the referenced documents and nationally published construction cost guidelines and available information on local market conditions. This estimate can provide guidance for your decision-making, however it is not intended to be a guarantee of the actual cost of a project. Frequent fluctuations in labor and building material costs, as well as the local bid climate, can substantially alter the construction cost of an individual project. The final determination of building cost is made through the bidding process with a contractor.

Following is a brief list of specific cost estimate preparation tips.

1. Work from the outside in—from larger systems to smaller components. Cover the major building components first (roof, siding/skin, doors and windows, sitework, expansive finishes, mechanical, electrical, and plumbing systems), then move on to individual elements (appliances, specialties, particular finishes).

2. Estimate systems, rather than individual components. For example, try to combine costs for a building's entire wall system rather than breaking it down into framing, insulation, cladding, and interior sheathing. This makes it easier to see the overall building costs and less likely that a major aspect of the work will be missed. It also makes it more convenient for the architect to compare costs between different schemes (if necessary). Standard cost estimating materials typically have component costs for different building superstructure (exterior enclosure) types.

3. Check the systems against Mean's Construction Cost Data® or other known databases to support the credibility of the estimate. Also, adjust the estimate to reflect known local conditions (union labor issues, spikes in material costs, or a difficult bid climate).

4. Use a contingency percentage appropriate to the level of detail of the documents and type of work on which the estimate is based. A renovation estimate based on schematic documents may require a 20–25 percent contingency; a new construction estimate based on full construction documents may need a 5 percent contingency. In general, the smaller the project the greater the uncertainty present in the estimate, and the higher the contingency factor should be as a result. Renovation work is always full of cost surprises.

5. Within the estimate, the architect should take care to use numbers that include subcontractor overhead and profit. Apply appropriate general contractor overhead and profit at the end of the estimate. Usually, this should be 10 percent overhead and 10 percent profit.

Client Management

It is better to be high on the estimate and lower on the actual bids. Clients get very upset when the actual bids come in well above the cost estimate. It happens all the time, of course, but it doesn't make it any easier to handle. When the estimate is high on the front-end, the worst that happens is the client picks the architect's estimate apart and claims he has overvalued the costs. An architect responds by reiterating this is his best assessment and recommends pulling out pieces of the work as possible deduct alternates in the bid documents. High early estimates make the process much more manageable and less stressful for everyone. Contractors use this technique all the time!

4

Bidding and Negotiation

Though often referred to as "master builders," few architects have the resources, capabilities, or inclination to actually construct the building they have designed. Therefore the owner must get someone to do it. The process of securing the person legally obligated to deliver the completed construction is commonly referred to as the bid process, or the bidding. The term derives from the practice of sending the construction documents to a number of potential contractors and inviting them to submit bids to build the work, with the lowest acceptable bidder usually getting the job. Although bidding is not the only way to obtain the services of a contractor, it is the most commonly used and the one most fraught with hazards for the design professional. Other methods of contractor procurement are mentioned in this book, but bidding the construction documents is the focus of this chapter.

No set of construction documents is perfect. Details the architect believes to be crystal clear are viewed by the contractor as confusing or unbuildable. Specification sections intended to tie down products are contradicted in the drawings or generate questions from suppliers. The engineer's sophisticated fire protection specifications cause a deluge of concerns: *"We don't think this will work; We've never seen anything like this before."*

Bidding and negotiation are those uniquely advantageous times in the project when the architect has an opportunity to include in the contractor's scope anything that he forgot, detailed poorly, or that was simply misunderstood by bidders. It is an opportunity to introduce more evidence to the court of bidders, to improve the case for what the contractor is obligated—and not obligated—to provide.

Architects too often look at bidding and negotiation as a burden, a sometimes unpleasant obligation they must meet to their client before turning their documents over to the contractor for building. In doing so, they underestimate the importance of this phase of services, and miss an opportunity to correct problems in their documents before the contractor is on board. The questions contractors ask during the pre-bid conference, and during bidding and the negotiation period, are all occasions to correct and to clarify. Anything that confuses a contractor during bidding will show up as either a higher cost in the bid or a dispute during construction. At no other time during the project does the architect have an opportunity such as this—almost a free pass—to have his documents critiqued by knowing professionals and an opportunity to revise them with impunity. Every issue or misunderstanding rectified during bidding is one less that will pop up during construction. It is sufficient to say that, next to taking the time to prepare quality construction documents, the bidding and negotiation phase represents the architect's most powerful tool in reducing his risk on a project.

SELECTING CONTRACTORS

Most architects know a small stable of contractors with whom they enjoy a good working relationship. They may even know of one particular contractor who has bid low on previous projects of a similar nature, who has a corral of capable and competitive subs, and who has impressed architects and owners alike with his professionalism and with his ability to deliver a constructed building on time and within budget. In a perfect world, this contractor would end up as the low bidder on their project. However, the vagaries of construction bidding means he often does not. As much as architects try to understand the bidding process, it often defies their comprehension, and makes the selection of contractors a vexing exercise seemingly conducted in fog and shadows.

In publicly-bid projects, the bid list is open to any contractor who can meet the statutory requirements for bidding. These generally consist of providing a bid and performance bond, non-collusion and affirmative action affidavits, and other requirements that open the field to a wide range of contractors. In these types of projects, the public entity is usually required to accept the lowest responsive bidder. This book deals with private construction, where the owner is normally free to select the bidders for his project based on who he believes will deliver the best value in terms of low bid, schedule conformance, and construction quality.

Some owners have a ready list of contractors they want to bid their project. Others will turn to the architect to develop such a list. In doing so, it is critical for the architect to remember he owes a financial (or fiduciary) obligation to the owner to suggest those contractors who are most likely to provide the desired benefits. This means that the architect's personal likes and dislikes should be suppressed in advising the owner on which contractors to add to his bid list. At the very least, however, the architect should disclose any biases, preferences, or affiliations with the potential bidders and note that he will conduct a fair bidding process on behalf of the owner. There are two main reasons for this.

1. A contractor whom the architect finds personally challenging and disagreeable may well be the most favorable candidate on a particular project. The architect owes his client advice on the best contractors for client's needs—not the architect's preference.

2. An architect's experience with a contractor on a particular project is usually the product of his interactions with the superintendent and project manager. His experience with different personnel on a new project may be poles apart from the earlier disagreeable experience.

Particularly with large projects, there may be a limited number of subcontractors in an area capable of properly staffing the job. No matter how many contractors an architect adds to the bid list, they end of trolling in the same pond of subcontractors.

Putting personal opinions aside does not mean that the architect must ignore all prior professional interactions with a contractor.

The architect's prior experiences with a contractor's management of the project are fair game to present to the owner, as long as they are presented in a balanced manner, fairly representing both the positives and negatives associated with working with that particular firm. A good way to present this information to an owner is through a confidential memo summarizing the strengths and weaknesses of potential bidders. The following is an example:

- Acme Construction: *Low bidder on two other similar projects; late with key submittals; construction quality very good; missed substantial completion date by two weeks.*

- Build-Rite Contractors: *Very schedule conscious—finished two weeks early on a similar project; higher change order percentage than most; two job site safety incidents on last project; construction quality good.*

- Fast & Good Builders: *Second lowest bidder on several recent projects, no recent work experience with them, good reputation, but little experience with this building type.*

This type of information, combined with input the owner may be seeking from other sources, can be invaluable to him in forming a mosaic of attributes for firms that represent the best candidates to bid on his project. It also creates a useful conversation with the owner over what to expect during the bidding and construction process. In the above example, if Acme is the successful low bidder the owner and architect know they need to focus on submittal submission as a key weakness of an otherwise strong contractor. This aspect should be discussed during contract negotiation, and all three parties should work to help the contractor improve in this area.

In helping the owner devise a bid list, the architect should keep the following tips in mind:

1. The bid list belongs to the owner. *The architect advises and counsels, but the decision belongs to the owner.*

2. Limit the list of bidders to five to seven firms. *This accomplishes the goal of letting every bidder know they have a reasonable opportunity to obtain the work, and it simplifies*

the bid process. The owner is generally not well served by a bloated bid list. As long as the firms on a shorter list are carefully selected, he will obtain a competitive range of bids with a shorter list.

3. Issue invitations and RSVPs: *As formal as this sounds, this is usually no more than a phone call. Each contractor should be told the owner is inviting him as a member of a limited bid list. He may ask, and should be told, which other firms are on the list. If the contractor declines to bid, it is usually for a very good reason (more work than he can handle; inability to compete with other bidders).*

4. Architect assessment of contractors: *False or malicious information stated by the architect about the contractor may be the basis of a trade libel suit against the architect if the owner relies on this information to deny the contractor the job. In discussing a contractor's suitability for a project with the owner, the architect should reference only his first-hand experiences, or the documented experiences of others, and avoid irrelevant or unsupported gossip and innuendo when commenting on a contractor's competence or integrity.*

THE PRE-BID CONFERENCE

Stockbrokers have a saying they use to describe a particularly bad day on Wall Street: "The market has spoken." The pre-bid conference is the architect's first opportunity to hear the bidder's market speak on the quality and completeness of his documents. See the Risk Reduction Tools for a sample pre-bid conference agenda. Experienced architects are often concerned when they hear a large number of fundamental questions from contractors at a pre-bid conference. This can simply mean that busy contractors have not spent much (or any) time reviewing the drawings and are verbally exploring them for the first time. It can also mean, however, that there is a basic lack of clarity in the information being communicated and it is confusing the bidders. Confusion during bidding is very bad, resulting in a wider range of bid values, bid exclusions and conditions, and potential scope disagreements during con-

Risk Reduction Tools

Sample Pre-Bid Conference Agenda

A. Circulate a sign-in sheet

B. Introduction of owner and professional team

C. Introduction of bidding contractors

D. Brief overview of the project

    1. Civil

    2. Architectural

    3. Structural

    4. Mechanical, plumbing, and electrical

E. Particular aspects of the project

F. Bidding procedures and deadlines

G. Questions from contractors and subcontractors

    Formal responses will be provided via bid addenda

H. Tour of site/property

    Submit questions generated by tour via bid RFIs

struction. To use the pre-bid conference effectively, the architect must prepare for it well and deliver an organized presentation that accomplishes the following basic goals:

- Schedule the pre-bid conference at least a week after contractors receive the bid documents. This allows the bidders to review the documents and develop a familiarity with them. The bid documents should never be distributed at the pre-bid conference.

- Introduce the owner and professional team: Contractors like to take the measure of who they will be working with on the project, so ask your principal consultants to attend the conference with you.

- Ask the contractors to introduce themselves: They are professionals as well, and are devoting time and resources to the project. Let them introduce themselves to the owner and the architectural/engineering team. Use a formal sign-in sheet to record their names and contact information.

- Provide a brief overview of the project: State the owner's goals and aspirations, the physical characteristics, and the general scope of the work.

- Note unique aspects: Clearly state any unusual or unique aspects of the work. Alert contractors to difficult working conditions or hazards.

- Take questions from the contractors: The architect should not answer questions at the pre-bid conference unless they are clear in the documents, nor should he attempt to answer any question in great detail unless he is confident in doing so. He should hear and record every question asked by the contractors for response later in a formal bid addenda.

- Tour the site or facility: It is very helpful to provide the entire group with a look at the existing site, or the building if the project is a renovation. The architect should not take questions during this period, but should try to highlight aspects of the work that he considers important.

- Architects differ on whether it is appropriate to keep minutes of the pre-bid conference for distribution to attendees, or whether it is best to keep the conference informal and summarize any information conveyed through formal bid addenda. Either way is appropriate, as long as verbal information conveyed by the architect or owner in the pre-bid conference is conveyed to the contractor in some means, either through pre-bid meeting minutes or bid addenda.

BIDDING TOOLKIT

Architects have tools available to them to help structure the bidding in a way that provides the bidders with a fair field on which to base their bids, and also gives the owner competitive options to meet his budget. To be effective, the use of these bidding techniques and strategies must be planned ahead of time and integrated into the construction documents. Owners and contractors must be

- Estimated Quantities

- Allowances

- Add and Deduct Alternates

- Unit Prices

- Phasing Information

- Alternate Substitution Bid

- Detailed Bid Response

FIGURE 4-1
The bidding toolkit.

made aware of their existence, why they are being employed, and how they will be used in the bid process. Unlike public bidding, the bid procedures for private projects allow tremendous leeway to structure the bid documents to meet the needs of the owner and to provide effective risk management for architects (see Figure 4-1).

Estimated Quantities

Architects often encounter situations, particularly in renovation, that will require work of some unknown amount. On a roofing replacement project, for example, an architect may know that a certain amount of decking under the existing roofing is deteriorated, but has no way of knowing the actual amount. One common response in bid documents is to toss the problem to the contractor with a specification or drawing instruction such as: *Contractor responsible for replacing all deteriorated decking*. The problem with such an

instruction is that it places a burden on the contractor to guess at
something that he can estimate no better than the architect. Rather
than lose a bid, he may guess low and make this assumption a
condition of his bid, complicating a comparison with other bids.
Furthermore, all the bidding contractors are forced to make their
own assumptions of the scope, potentially resulting in a wider bid
variance and more money eventually paid out by the owner if
every bidder has assumed the worst.

The solution? Architects can use estimated quantities to define a
part of the project that cannot otherwise be quantified. In the roof
decking example, the architect could estimate that 600 square feet
of decking would require replacement. In the construction documents, he would specify the type and quality of decking (presum-
ably to match existing). The contractor would then be obligated to
include this quantity of decking replacement in his base bid. He
should have no problem in doing so, since he now has a defined
scope on which to bid, and he is on a par with all the other bidders. How did the architect determine the amount of the estimated
quantity? He guessed. It was an educated guess, and may have had
some basis in his examination of the existing roofing, prior experience with similar projects, and his conversations with roofers who
had performed other work of this type. Still, it was an estimate
created for the benefit of the owner to quantify an unknown item.

Estimated Quantities and Unit Prices

So what happens when the roofing work is completed and the
actual amount of decking replacement is quantifiable? Obviously,
the total amount is very unlikely to have been 600 square feet,
given the speculative manner in which that figure was determined.
But this discrepancy does not matter.

Why? The architect coupled the estimated quantity with a unit
price. A unit price provides a means for the contractor to provide
a *bid* value for additional work required if the estimated quantity is
low. For instance, if the required deck replacement actually totals
700 square feet, the contractor would be paid for an additional 100
square feet of deck replacement at the unit price he provided in the
bid documents. Conversely, if the actual deck replacement quantity

turns out to be 500 square feet, the contractor will owe the owner a *credit* for 100 square feet of decking that the contractor assumed in his base bid but did not have to perform. Once again, this credit would be calculated based on the unit price value the contractor indicated in his bid.

Coupling estimated prices and unit prices in the bid documents provides the advantage of building the bulk of the cost of some unknown repair into the base bid, and covering the rest with a unit cost, so the contractor is encouraged to make it as competitive as possible. The fact that the contractor may end up paying the unit price back to the owner as a credit is further inducement for him to make the unit price responsible. Architects often can build in the following two additional advantages for the benefit of the owner:

- Specify in the bid documents that all unit prices are nego-tiable. If a contractor submits a unit price that is clearly excessive, the owner can correct this during contract negotiation. It is usually sufficient to point out to the contractor that his unit price may be used to calculate a credit, and to share other unit prices submitted by competing bidders, to convince him to revise his unit price to an acceptable value.

- Owners like credits better than extras. Realizing this, clever architects try to set their estimated quantities at conservatively high numbers to ensure that the contractor will owe a credit to the owner versus an added cost. This strategy can backfire on the architect, however, if he is too aggressive in building credits via estimated quantities into the documents.

One problem with using estimated quantities is documenting the amount of work that has actually been performed. In the example of roofing replacement, this might be easily determined by a visual inspection if the architect was able to visit the site often enough to track the work. In other instances of lesser work (e.g., the replacement of deteriorated wood joists) the architect may not be able to view the extent of work before it is covered up. Although some independent documentation (material delivery job tickets) may be available, in these instances the owner and architect are mostly required to rely on the integrity of the contractor.

Allowances

Allowances can be a useful tool in providing value in the bid documents for finish items that the owner has not yet selected, or for small items that have not been detailed or specified, yet the owner wishes to include. Examples of finish items include brick, tile, stone, or other bulk finishes that need to be included as part of the bid but have not yet been selected. In consultation with suppliers, the architect should try to set a reasonable allowance that covers a wide range of finish selections. The architect likely has some idea of the quality and colors the owner will be looking for, so a conversation with a supplier should give him a safe value to insert as an allowance.

Since finishes with extensive areas can be quite costly, the architect should take some care in setting allowances. Bulk finishes are bid items, versus purchased items, so using an allowance removes the benefit to the owner of the competitive bid climate for the value of the finishes. Many architects have observed the tendency of the actual selected finish to be very close to the allowance value. Rarely, it seems, does the value fall well below the allowance. This may not be totally due to the machinations of suppliers. It may also be attributed to the basic nature of people. Given an *allowance* to spend, owners will tend to use it all. Being the contradictory creatures they are, owners will use up their entire allowance on finishes with no thought of saving money while still complaining that the architect designed a building that was too expensive. The lesson for architects? Because of their impact on the budget, select bulk finish allowances carefully. Too high and the architect can bust the budget. Too low and the owner will be frustrated with the narrow range of selections.

Setting allowances for small items, such as ornamental plaques, benches, or cornerstones simply requires finding a recent price for similar items and installing it as the allowance. Since these stand-alone values are smaller, and easily documented by the architect, there is very little reason not to guess high on these allowances. Architects should be wary of using allowances for more extensive work or items with multiple installations. Any piece of the project that constitutes a large value should be determined ahead of time and included in the construction documents.

Add and Deduct Alternates

Add and deduct alternates are a very useful tool which allow the architect and owner to use the marketplace of bidders to help them assess how much scope they can add to or subtract from the project to best utilize their budget. Although both are similar in nature, add and deduct alternates have some fundamental differences beyond the obvious.

One warning: Contractors will often ask during the bid period for a prioritized list of which alternates the owner will accept first. Such a list is often required for publicly-bid projects but is optional for private projects. Contractors are concerned about this issue for two reasons: 1) They want to ensure they are competitive on the add alternates the owner is most likely to select, and 2) They are concerned that the owner and architect will use selective groupings of alternates to slant the bid toward the contractor they truly want to obtain the work. In private sector work, this second issue is a red herring, of course, since the owner could simply award the work to a contractor of his choosing, or award the work to a higher bidder if he so chooses. Architects and owners can decide if they want to provide bidders with a prioritized list of alternates. In general, there is little reason to do so since the likely result would be for the top one or two alternates to be priced very competitively while the remainder may vary wildly.

Add Alternates

Add alternates should be employed for only those aspects of the project that the owner considers desirable, but not essential. Examples could include the installation of an exercise center at an industrial facility or an upgrade from ceramic tile to marble tile in an office lobby.

Another reason architects may use add alternates is to keep a supplier from offering the owner less than competitive prices. Some products are specifically selected for a project by the architect because he believes they best serve the use and the owner's interest. In the instance of a product such as windows or glazing, a large amount of coordination between owner and supplier may take place prior to bidding, with the supplier devoting time and energies to working with the architect to provide him with technical guidance, specification information, and standard details. The product representative does so with the full expectation that his product will be used on the building, and for his part, the architect produces details and specifications focused on this single product. All of this is in the owner's best interest up to the point of bidding. At that point the profit motive of any reasonable businessperson may take over if he recognizes that he is the only specified provider of windows on the project. Contractors complain bitterly about dealing with sole-source providers for any component of a project due to its impact on their ability to obtain competitive pricing, and in general their complaints are understandable. Architects usually address this situation by listing other products as acceptable alternates in the specifications, or allowing "or equal" products to be proposed by the contractor and judged by the architect at a later date. The problem with this approach is that the architect loses control of the specification. Having devoted considerable time and energy to selecting and detailing around a particular product, he throws it all out the window in the name of competitiveness. A better option may be to set up an add/deduct alternate for another product of his choosing that will provide similar quality, as well as the desired competition for the most favored product. The architect may well include a couple of standard details to support the pricing of the alternate, and indicate to all that he is well prepared to accept the alternate if necessary.

Where add alternates are incorporated into the construction documents (and if they are not incorporated into the documents—how is the contractor pricing them?), they need to be clearly identified as alternates to avoid having rushed subcontractors or suppliers including them in their base bid numbers to the contractor. It also simplifies the contractor's job in separating the base bid scope from the alternate scope for all his subs and suppliers.

One way to facilitate this process is for the architect to include a separate line on the bid proposal form for each add or deduct alternate. The architect should also include a separate specification section for add and deduct alternates, describing in narrative form the full extent of each.

Deduct Alternates

Deduct alternates are a way for the architect to build a form of controlled cost-cutting into his construction documents. Deduct alternates are discrete areas of the project carved out and set aside as stand-alone objects to be priced by the contractor. If the lowest bid for the overall project is beyond the owner's budget, the architect can help bring the project back within the budget by helping him to select one or more deduct alternates to reduce the costs. Typically, deduct alternates come in the following two flavors:

1. ***Non-essential, stand-alone items:*** These are items that are desired by the owner, but can be removed in total with no serious detrimental effects to the project. Examples include: play structures, fencing, emergency generator, wainscot, or crown molding.

2. ***Alternate items:*** Alternate products or construction elements to those specified in the base bid. These normally revolve around finish options, where the architect sees an opportunity to either save the owner money through less-expensive options, or to ensure that the preferred finish provider is more responsible in his bid by setting up a competition with other lower-cost providers. Examples include: carpeting and other flooring products, windows and glazing, and exterior cladding products. Cost-cutting by substituting less expensive materials and systems without cutting the scope is sometimes referred to as *value-engineering.*

In general, contractors despise deduct alternates. They view them as a form of devious manipulation of the bid by the architect, and a threat to their goal of winning the project outright through the lowest bid for the full scope. As a result, they tend to respond by discounting deduct alternates below their true value, reasoning that the owner wants to build the full project, and consequently does not want to exercise a deduct alternate to remove scope. They believe they will win the competitive bid, and do not intend to give away much beyond that. For this reason, deduct alternates only have value where the architect believes the bids will be very close to the owner's construction budget, and feels that the owner's budget may require some minor relief. If the architect is concerned that the bids may be well beyond the budget, deduct alternates are not a desirable tool to use in correcting the situation. Deletion of scope from the base bid is the more desirable route, with the option of an add alternate to price out the cost of restoring it.

Unit Prices

Unit prices are specific prices requested from the contractor for additional work of a limited scope. Use of a unit price assumes that the scope cannot be defined, otherwise the architect would include it in the construction documents and take advantage of the bidding environment to obtain the best value for his client. Unit prices are set up on a *per unit* basis, meaning that the contractor is asked to provide a total value (including labor, materials, overhead, and profit) to a single unit of the item. Unit prices are used for such items as additional concrete paving, additional brick veneer, additional sod, or additional fencing. They are most often applied to site items, though they can be used for practically any component of a project where the owner believes he may want to add some incremental scope but has no idea how much. One problem with unit prices is that when they are not coupled with estimated quantities, the contractor is more likely to include an extremely high figure. Another problem is that with a unit price the contractor has little information on which to base a competitive price. The unit price may be employed for a very small quantity or a very large quantity. For this reason alone, a contractor's unit prices are typically not very competitive and provide little value to the owner except that of predictability. Even with these facts, the architect should state clearly in the bid proposal that unit prices are not considered in judging bids and are subject to negotiation. This provides the owner the opportunity to negotiate down unit prices that are abusively high.

Phasing Information

Phasing of a project occurs when site, occupancy, or monetary restrictions require the contractor to divide the work into separate stages, with each stage preceding the next. One of the most vexing experiences of architecture is to be sitting in the first job meeting to discuss construction of a project and simultaneously realizing that phasing will be required to meet the owner's needs, and that the contractor did not factor this into his bid. Time, in the form of a contractor's general conditions, is a major cost factor in any project. When an owner's circumstances require a project to be implemented in phases, there can be severe cost implications for the project and equally severe embarrassment for the architect who did not include the necessity of phasing as part of the bid documents. In cases where existing commercial, residential, or industrial facilities need to remain operating during construction, architects must always discuss with the owner basic phasing and include as much information as possible in the construction documents for the contractor to consider in his bid assumptions. It can be very difficult to determine details of phasing during the construction documents phase of a project. An owner himself may not realize all the implications of the work he is undertaking, and contractors add their own expertise once they enter the project. Nevertheless, it is necessary for the architect and owner to work out basic phasing, not tied to a timeline, to alert the contractor to how the project will be structured. This information should be as detailed as practical, but may be little more than a key plan with notes showing how each phase is organized. Even this small amount of information lets the contractor understand the nature of the project and include provisions for phasing the work in his bid. Obviously, the more information provided to the contractor, the better he can respond to the owner's needs.

Alternate Substitution Bid

Contractors and their subs like substitutions. They like them because they provide the opportunity to undercut the competition and win the bid. They like them because they provide a way to tap into extra profits. They like them most of all because they are the American way: Build a better mousetrap (or at least a less expensive one) and the world will beat a path to your doorstep. Although most architects would argue that they really want the specified product on their job, they understand that substitutions are a necessity to ensure competitive bidding across the board, and to allow for the possibility that an equivalent product may be better than the specified product (or at least available if the specified product is not). Owners benefit from substitutions because they obtain the best value for their money. This entire process normally works well, as long as it follows the program: Architect specifies the product he wants; architect writes a specification listing an acceptable product that is equal to the one he specified; contractor bids to all and selects the subcontractor with the lowest-cost acceptable product. But what happens when, competitive fires burning brightly, the contractor finds another product not listed by the architect that he considers an equal, and includes it in his bid? If he wins the bid, this inclusion may not be noticed until later in the project, when the architect receives the submittal and must either reject it (and likely battle with the contractor or his sub) or accept it. It is important for the architect to stress to all contractors that any substituted product not listed as an equal in the specifications must be proposed in a formal substitution request, and not arrive on the architect's desk in the form of a submittal. See Chapter 6 for a more thorough discussion of substitution requests.

An alternative method in private bidding is to provide the contractor with a ready means to throw in every substitution and cost savings measure he can conceive—and thereby win the project in the process. The alternate substitution bid allows contractors an owner-sanctioned means for contractors to use their contacts and subs to show the owner a less expensive way to achieve his goals. The advantages to this system is, if bidders cooperate, it pushes all the hidden substitutions into the open and gives the owner another option to constructing his building if all the base bids come in over his budget. The downside is that it creates a wild card of sorts, where a contractor can propose a major systems change that is far inferior to the specified systems, not workable at all for the owner's needs, and still have it seriously considered during the bidding. Still, the alternate substitution bid can be useful as a fail-safe. When all the bids are opened and over budget, the architect and owner are left with two options: Look for cost savings, revise the drawings, and rebid; or sit down with the lowest-bidding contractor and negotiate. The alternate substitution bid gives the owner more options by providing him with contractor suggestions of ways he can save money. Upon examination, many of the proposed substitutions may prove to be totally unworkable for the owner. Others, however, may be useful and may move the architect and owner toward a workable solution to their budget problem.

Detailed Bid Response

Architects may include in their bid proposal form a request for the contractor to provide a detailed bid response in his proposal. Though more often seen in public bids, this request is gaining favor in private bidding as well because it provides the owner with basic costs for each major division of the work. The architect and owner can compare this information to previous cost estimates to determine why the project is over budget, or to assess how well prior cost-saving measures worked. By examining this breakdown across contractors, the architect may be able to discern where an unreasonably low bidder missed significant scope, or in the case of widely divergent bids, where his documents may have lacked sufficient detail.

See the Risk Reduction Tools for a list of bid tips.

Risk Reduction Tools

Bid Response Form Tips

- Make sure the bid response form requires the bidder to state the addenda he received.
- State the bid alternate numbers and titles on bid response form; provide a separate line for each alternate.
- Require the bidder to clearly note ADD or DEDUCT for each alternate.
- Provide separate areas on the bid response form for unit prices. Where a unit price is tied to an estimated quantity, state this clearly in the bid response form.
- If allowing a contractor to propose substitutions, restrict this information to an alternate bid response form. Do not try to include substitutions on the standard bid response form.
- If requesting a schedule duration estimate on the bid form, define whether duration is calendar or week days.

QUESTIONS DURING BIDDING

Contractors submit questions during bidding because they are trying to help clarify unclear aspects of the construction documents, and because subs and suppliers pass questions through them for the same reason. Bid questions take all forms, in a wide range from foolish to disturbing. Simple questions can be easily answered by referring to the appropriate detail or specification section. Other questions at the opposite end of the spectrum can raise serious concerns in the mind of the architect as to whether he or his consultants have made a serious misjudgment in the construction documents. More often, however, bidders are likely to catch the common assortment of coordination errors, mis-keyed details, schedule mistakes, dimensional oddities, or other small mistakes that often result from completing the documents under a tight schedule. Arguably, these types of errors should be caught by a thorough in-house review prior to issuance of the bid documents. In busy practices, however, such a review is a rarity. So when contractors raise these items that architects may consider *clerical* in nature, they should be thankful. Correcting the record during bidding by providing a missing detail, clearing up a mis-stated drawing scale, or making any number of other small clarifications can save real trouble and real money once the contracts are signed and construction is underway. Small clerical anomalies during construction can be painful and expensive to explain to owners, particularly when there is a change order attached to them. Bid questions are a free review of the documents by hundreds of trained eyes. An architect who takes the time to correct every flaw—large or small—that he finds during bidding saves himself much trouble down the road. A few bid question tips:

- Maintain the integrity of the bid process: *Insist that all bid questions be submitted as requests for information (RFIs). Answer promptly, respond formally in a written addendum, and send responses to all bidders at the same time. In short, the architect must work to keep the playing field level for all bidders.*

- Set a deadline for questions: *State in the bid documents the last date on which bid RFIs can be submitted, and hold firm to that date. Usually a week ahead of the bid's due date is reasonable, in that it allows a day or so to respond and gives the contractors a sufficient amount of time to distribute the RFI responses to their subs and suppliers.*

- Answer questions fully and precisely: *When a bidder asks a question, it is because he is confused about what the documents are requiring of him. This confusion is detrimental to the architect and the owner, and clarifying it is essential to helping to promote the most favorable bid results for the owner. Answer questions completely. If the contractor has simply missed a detail, refer him to the detail and specification section that applies. If a contractor is stating that information is missing or incomplete, respond with additional information if he is correct. If the architect does not feel the information is deficient, briefly recap the pieces of the drawings and specifications that describe the work he is questioning. Overall, the architect should maintain a professional and supportive tone in his responses. Contractors are assessing the situation to see if they will be dealing with an architect and owner who are reasonable, competent, and cooperative.*

BID ADDENDA

Bid addenda are an opportunity to revise the record. During bidding, the architect will benefit from having a diverse collection of skilled eyes looking over the documents for the first time. Like a person reading a novel, bidders will find some sections lacking and some too full of detail that doesn't matter to them. Each will bring his own prior experiences and prejudices to reviewing the documents, and will respond based on that history. Where these questions or concerns affect the bid, the contractor will seek to have them clarified by the architect through submitting a bid request for information (RFI). Bidding contractors also distribute the drawings to a number of subcontractors and suppliers, and will receive questions from them about the documents that they will pass on to the architect. See the Risk Hazard Flags for list of bid addenda traps.

Risk Hazard Flags

Bid Addenda Traps

Bid addenda are used to answer bidder questions and fill in gaps. Avoid these addenda traps:

- Responding vaguely or glibly
- Not referencing the construction documents when answering questions
- Not including a missing detail or specification section in the addenda
- Expanding the contractor's scope without additional information
- Adding owner responsibility without his consent
- Not sending addenda to all bidders in the same manner and same time
- Issuing addenda a few days before the bid date

NEGOTIATING THE CONSTRUCTION CONTRACT

Negotiation of the construction contract represents a unique time in the life of a project. In the best case, the owner has a low bidder ready to start the project. Even if his bid is below the owner's construction budget, there are other reasons to negotiate with the contractor. The most important of these reasons is to determine if the low bidder is, in fact, the *lowest responsible bidder.* Architects tend to think that bid responsibility results from a bidding contractor submitting a set of properly completed bid documents. Determining bid responsibility is much more than that. It consists of *descoping,* or reviewing with the contractor the scope and some of the intricacies of the project to determine if his low bid fully represents the scope, or if he has, in the parlance of the trade, "left money on the table."

The period before signing of the contract is the most advantageous time to negotiate changes to the bid, flush out missing or redundant requirements, review add and deduct alternates, and explore opportunities for cost savings.

Normal negotiation topics include: Value-engineering opportunities identified by the contractor or his subs during bidding; non-competitive unit prices; potential change orders from gaps or incomplete information in the construction documents.

Owners need to be prepared to add cost to the contract to avoid change order or schedule delays later resulting from contractor interpretation revealed during negotiation that differs from the architect's intent. Bids that are substantially below the second lowest bid need to be reviewed carefully to ensure scope is not missing. Phasing limitations, special conditions, and other project concerns need to be discussed during negotiation to ensure the Contractor is aware they exist, and has included them in his bid.

Labor concerns and strategies need to be discussed, with common understanding among all parties.

The owner needs to review the list of subcontractors proposed by the contractor. He has the right to reject those he deems unacceptable, but may have to pay for the privilege.

Scope that was omitted from the construction documents may be added if it can be clearly documented and quickly priced by the contractor.

The owner and contractor should have a frank discussion regarding schedule expectations or limitations. Bonus clauses may be added during negotiation. Penalty clauses may not.

During this portion of the negotiation, the architect should ask the contractor such questions as:

- Did he include provisions for phasing, temporary fencing, or security personnel?
- Did he factor into his bid the estimated quantities and allowances included in the bid documents?
- How many months of general conditions has he budgeted for?
- Did he include provisions for field engineering and testing required in the specifications?
- Is he aware of unique aspects of the project?

These are only a few general questions that an architect may want to ask a contractor to ensure that his bid, whether competitive with other bidders or disturbingly below the next lowest bid, is valid. All of the responses are documented and included in the contract for construction. See Figure 4-2 for a list of potential items that are referenced in the contract.

Another reason to negotiate with the low bidder is to carve out some breathing room in the budget for the owner. Bids have an annoying way of pushing up against an owner's budget, and it is often necessary to negotiate with the contractor to create some

FIGURE 4-2
The contract for construction.

budget flexibility to allow for additional contingencies and unexpected change orders. Hopefully, the construction budget already has a contingency factor built in, but if it does not there is all the more reason to work with the contractor to look for savings.

The advantage of doing this before the contract is signed is obvious. Once the bidder signs the contract and becomes the contractor, credits have a way of decreasing in value to below what they would have been in a more competitive environment. During negotiation, the contractor is fully aware that his ability to reel in the owner and get the project under contract is dependent on his ability to help the architect and owner find sufficient savings to make the project viable. Yes, he realizes he is the low bidder. He may even know how much lower his bid is than the next highest bid, but he does not know the owner's true budget and is aware that the owner may be negotiating with another contractor as well. For these reasons, negotiation before the signing of the owner/contractor agreement is the ideal time to get "while the getting is good." It is the ideal time to verify the contractor's understanding of the project scope and to seek out any readily available cost savings that he or his subcontractors identified through bidding, or that they can offer up as a way to obtain the contract with the owner.

BONDS

Bonds refer to a small collection of insurance-like products that protect owners from contractors who are unable to fulfill their contractual obligations, or provide municipalities with assurance that a property owner will fulfill his obligations. A rule of thumb regarding the differences between bonds and insurance: insurance covers things that happen, while bonds cover things that do not happen. Payment and performance bonds are desirable to clients on private projects as a means of ensuring they have a more stable (bondable) contractor, and as assurance that their project is not imperiled by a potential default of a contractor.

Much of the risk of a contractor default can be removed through prudent use of the tools normally available in construction administration: Selection of a contractor with a reliable and long history of work, schedule of values that fairly reflects the value of each phase of the work, 10 percent retainage, architect's involvement in assessing the application for payment against the work in place, monthly applications for payments, diligent use of partial lien releases from subs and suppliers.

Typical bonds encountered in construction include:

- *Bid bond:* This bond is usually required in public bidding, and consists of a bond purchased to cover a percentage of the contractor's bid (usually 10 percent), to ensure that the bid is responsible and valid for the period required in the bid documents. Bid bonds are rarely used in privately-bid work.

- *Payment and performance bond:* A combination bond that obligates a bonding company to take over the work and complete it for the original contract value if the contractor defaults for any reason. These types of bonds are used heavily in public work, and to a lesser extent on private projects as well.

- *Maintenance bond:* A bond required of a property owner by a municipality to ensure that the owner maintains the property as approved by the municipality for a specified period (usually a year). The purpose of this bond is to ensure that landscaping is properly maintained and well established for appearance, as well as soil erosion and sedimentation control purposes.

- *Warranty bond:* A bond provided by the contractor to ensure that if he defaults in warranty obligations, a bonding company will arrange for other firms to provide the required warranty repairs for the specified warranty period (usually one year). This type of bond is normally used only on projects with sophisticated or extensive HVAC or other systems for which significant warranty work is anticipated.

DEALING WITH LABOR UNIONS

Managing the labor on a project is the responsibility of the contractor. Labor unions are prevalent in many places, particularly urban areas, and represent the skilled trades in negotiating for benefits, higher wages, and better working conditions for their members. The general contractor is responsible for dealing with the labor unions on any construction project he undertakes. Generally, the architect should have only occasional dealings with labor unions, but there are several situations where the owner looks to the architect for advice and expertise regarding project labor. During the bidding of a project, the bidding contractors will also be expecting the architect and owner to clearly define whether the project is union or *open-shop*. Open-shop projects are those in which the contractor is free to use union or non-union labor. Owners will often press for their project to be built with non-union labor because it is less expensive. Cost estimators generally assume union labor is 10 to 30 percent more expensive than non-union labor. Labor unions will argue that the productivity and professionalism of their members more than offsets this excess cost. In urban areas especially, the larger mechanical, electrical, and plumbing subcontractors tend to be unionized. Because of their size, and their ability to easily pull in extra labor from the union halls, these subcontractors can often staff a schedule-intensive project more readily than their non-union counterparts. Where double-shifts and weekend work are required to maintain schedule, union labor (for premium wages) is usually much more available and amenable to working overtime.

Owners are primarily interested in the bottom line, however, and will seek to have their project built with open-shop labor wherever possible. This presents an interesting challenge during bidding for the architect. While some owners are adamant that their project will be bid upon as an open-shop, others seek the architect's guidance in structuring the bid documents to both preserve labor peace on their project while also obtaining the most favorable bids. Bidding contractors want definition in the bid documents. They will ask the architect to define whether the project is all union or completely open-shop because they want to know how level the playing field is among bidders. Architects should resist this demand for labor clarity from bidders with a simple response: *The contractor is responsible for, and he possesses the expertise in, managing the labor on the construction site. The owner seeks a proposal utilizing the most favorable mix of labor for both low costs and construction site harmony.*

The irony is that construction projects in this country can, and often do, consist of both union and non-union labor. Contractors in heavily unionize areas are particularly adept at *mixing and matching* labor to provide the best advantage to the owner. Labor unions will sometimes *buy down*, or subsidize, a union subcontractor's bid to match a competing non-union bid. In other instances, labor unions will agree to accept other open-shop subcontractors on the project as long as their union is represented at some agreed upon percentage. Some open-shop subcontractors, particularly in the mechanical, plumbing, and electrical trades, operate both union and non-union sides of their business to take advantage of as many construction opportunities as possible. Even open-shop subcontractors on a predominantly union project can negotiate with some unions to pay union wages and benefits for their workers for the duration of the project.

All of these examples indicate that savvy general contractors have many tools available to them to craft a proposal for the owner that represents the best combination of labor costs and job site peace. When they fail in this goal, the result can be problems both on and off the construction site. Labor unions may picket the entrance to the site or use other public relations tactics to create public ill will for the owner and persuade him to accept union labor. Delivery services or other suppliers represented by unions may refuse to cross picket lines, increasing costs and causing schedule difficulties for the contractor. Instances of vandalism, anonymous violation reports to building code officials and regulatory authorities, and increased job site theft have also been reported as side effects of situations in which the contractor could not preserve job site labor harmony. Such instances are not typical, are illegal, and are not condoned by the national labor organizations. They do occur, however, and represent real costs to the owner, which must be considered in any decision regarding the mix of labor on a construction site.

BID EXCLUSIONS AND CONDITIONS

Bidders will sometimes include a short list of exclusions or conditions with their bid proposal. Bid exclusions are parts of the scope of work, either represented in the documents or not, that the general contractor is excluding from his proposal. Conditions are limitations or modifications that the contractor includes as part of his bid proposal. Sometimes these conditions are not in conflict with the documents. Other times, they can fundamentally alter the intent of the documents. (See the Risk Hazard Flags for examples of bid exlusions and conditions.)

Bid exclusions and conditions can occur for one of several reasons, including:

- Restatement of construction document information: *If the construction documents state that the contractor has no responsibility for remediation of a known environmental problem, the bidder may want to restate this fact as an exclusion to his bid to ensure there is no misunderstanding of a major issue.*

- Exclusion or conditions resulting from an unclear scope: *If a bidder feels some aspect of the scope of work is so unclear that he cannot reliably bid on its cost, he may either exclude this work from his bid or place conditions on it so as to define it in a way that allows him to reliably bid its cost. Example: "The bid is conditioned on the parapet fascia being clad in stock metal cladding versus custom fabricated material."*

 Risk Hazard Flags

Bid Exclusions and Conditions Examples

Bid exclusions and conditions make it difficult to compare bids among contractors. Watch for these warning signs in bid responses:

Exclusions
- Scope is excluded from bid
- Scope is proposed to be performed on a time and materials basis
- Scope is assumed to be furnished by "owner or others"

Conditions
- Scope is limited by an allowance
- Scope is altered in nature
- Scope is contingent on action by owner or architect

- Exclusion of an unknown problem: *If the bidder has identified a problem that is either unknown to the owner or architect, or not addressed in the construction documents, he will exclude it from his bid. Example: "The bid does not include any required remediation of pigeon guano from the attic spaces."*

- Exclusion or condition for competitive advantage: *Though not generally a wise tactic, contractors will sometimes use selective bid exclusions or conditions as a way to gain a competitive advantage in their bid. This technique often takes the form of limiting the cost of a portion of the scope of work, or proposing substitutions that provide them with a cost advantage over other bidders. Examples: 1) "The bid is conditioned on an allowance of $10,000 for removal of unsuitable soil areas designated in the geotechnical survey," or 2) This bid is conditioned on the architect's acceptance of Free Air HVAC equipment in lieu of the specified Rarified Air HVAC equipment."*

Architects and owners generally despise bid conditions and exclusions. They complicate comparing bids, communicate an air of sneakiness on the part of the contractor, and annoy owners who are not sure whether their architect's documents are deficient or whether they have a very persnickety bidder. For these reasons, contractors tend to avoid including any conditions or exclusions with their bid. They know that owners view them as an early indication that a contractor is likely to be tedious and difficult to work with. Even if a contractor harbors real concerns about the scope, or bases his bid on the presumption of obtaining a substitution for a significant specified product, he is more likely to reveal this in contract negotiations or later during construction when the owner is more amenable to working out a solution.

Some architects include stipulations on the bid form that state "no exclusions or conditions will be accepted." These types of statements are generally not effective, since a contractor who genuinely feels an exclusion is required to reduce his exposure is not going to be dissuaded by a prohibitive statement. He knows that if he is the low bidder, his bid will be closely examined (exclusion and all), and that if he is not the low bidder his exclusion does not really affect the situation.

An architect reviewing a low bid that contains significant exclusions or conditions faces a difficult dilemma. He cannot fairly assign a value to either one since he cannot know the true cost or savings to the contractor. Even under the looser guidelines of private bidding, he cannot go back to the contractor and ask him to revise his bid without conditions. The contractor likely had significant cause, at least in his own mind, to include the exclusions or conditions, and is probably not willing to remove them without additional clarification from the owner or architect. The architect is left, then, with two possible courses of action in advising the owner when the low bid contains exclusions and conditions:

1. Advise the owner that the apparent low bid contains significant exclusions and conditions that make it impossible to compare it to other bids without these items. It may be preferable to negotiate with the next lowest bidder, whose bid does not contain exclusions or conditions.

2. Advise the owner that the apparent low bid is significantly below other bidders, but contains exclusions and conditions of unknown value that must be negotiated.

LIQUIDATED DAMAGES AND BONUS CLAUSES

Liquidated damages (LD) and bonus clauses are tools used by owners to encourage the contractor to complete a project by a certain date for the benefit of the owner. While a bonus is unambiguously used for the purpose of encouraging a contractor to finish before a pre-determined date, this is not the legally permissible purpose of liquidated damages. With liquidated damages, the real purpose is to establish by mutual agreement between the owner and contractor an amount that will define the monetary injury suffered by the owner if the completion of the project is inexcusably delayed. Liquidated damages exist because it may be difficult for either side to precisely determine the actual damages when they occur, so they simply agree on an amount beforehand. If a court of law decides later that the daily liquidated damages amount is excessive or meant to intimidate the contractor, it will be rejected as a "penalty," and not enforced. For this reason, architects should ask owners and attorneys to carefully consider the per diem liquidated damages amount they stipulate in the bid documents.

Usually, though not always, the time allowed for construction is shorter than normal and the owner faces financial penalties of his own if the work is completed after this deadline. Bonus clauses provide for an increase in compensation for the contractor if he completes the work by this date. It is usually left up to the contractor whether he passes some of this bonus to his subcontractors to encourage them to meet the deadline. Bonus clauses are the carrots of the construction industry.

Owners tend to eschew bonus clauses in favor of the liquidated damages clause, a dollar amount charged per calendar day to the contractor if he exceeds the deadline. LD clauses have the following two effects that owners do not often count on:

1. Contractors have a fairly good idea of how long the construction should take. In bidding, they usually calculate the liquidated damages for a normal construction duration and add it to their bid amount. If they can pull off the construction in less time than normal, the LD becomes a bonus in their eyes.

2. In projects with liquidated damage clauses, contractors document any potential delay in the project and attempt to offset the LD by claiming additional days of construction are due to them. Claims for additional time include adverse weather conditions, architect's delay in submittal and RFI responses, owner decision delays, change order work, owner's own forces delays, owner-purchased equipment delivery delays, or municipal inspection delays. There is no shortage of reasons why a contractor will claim delays when faced with a liquidated damage clause, and the result can be an owner and contractor standing at the end of the project with competing claims in hand.

   Architects should advise owners, where possible, to allow sufficient construction time to avoid the need to assess liquidated damages against a contractor. They are useful as a means of reinforcing the owner's deadline, but a prudent owner will often employ them in the contract—but rarely in practice. If sufficient time is available to construct the facility and the owner feels more comfortable with an LD clause in place, it will do little harm. If the construction schedule is squeezed, the architect should try to persuade the owner to use a bonus clause since it is more likely to encourage the contractor to complete the facility under a deadline than will the liquidated damages clause.

5

Under Construction—The Basics

STARTING THE JOB

The architect has a number of responsibilities on behalf of the owner at the beginning of a construction project. While most of these may seem clerical in nature, they are critical pieces of the architect's responsibility to protect the owner's interest. Once the agreement between the owner and contractor is signed, the architect should obtain a copy for his use throughout the project. As discussed in Chapter 2, the architect may have been involved in reviewing this agreement since various provisions relate to his service to the owner in reviewing submittals, applications for payment, and requests for information. See the Risk Reduction Tools for a job startup checklist.

The architect not only has an interest in the contents of the owner-contractor agreement that affect his obligations, but in the attachments as well. During the course of bidding and negotiation, numerous documents may have been created that represent a part of the contract for construction. These documents include:

- Pre-bid conference meeting minutes (or notes of pre-bid conference discussion distributed through bid addenda)
- Bid addenda
- Bidder question or bidder request for information responses (if not captured in a formal bid addenda)

Risk Reduction Tools

Starting the Job Checklist

- Owner/contractor agreement signed
- Construction schedule complete
- Schedule of values prepared and accepted
- List of major subcontractors received
- Insurance certificates received
- Bonds (if necessary) received
- Pre-construction conference scheduled
- Notice to proceed issued
- Permits granted

- Memorandum of owner-contractor negotiation
- Architect's response to building permit review comments (should be captured in formal bid addenda, if possible)
- Schedule or construction duration agreements
- Schedule of value or retainage reduction agreements
- Any other document that modifies the bid documents.

Once the agreement is signed, the contractor has obligations to provide a number of documents to the owner before he commences construction. These may be sent to the architect for review. If they are acceptable, he forwards them to the owner for his approval. These documents include:

1. ***Certificates of insurance:*** builder's risk; general liability; automotive liability; and worker's compensation. All insurance certificates, properly endorsed, should be received before the contractor occupies the site or commences work.

2. ***Bonds:*** If payment and performance or material and labor bonds are required, they should be forwarded by the contractor before commencing work.

3. ***Schedule of values:*** The contractor should prepare the schedule of values that assigns portions of the contract amount to each major line item of the work. This schedule will accompany each application for payment.

4. ***Construction schedule:*** A critical path schedule showing line items for each major portion of the work (typically the same items shown on the schedule of values), their duration, and relationship in the progress of the work. This schedule should indicate the overall construction time, and indicate a milestone date for substantial completion.

One additional note regarding the construction schedule: Often, the contractor will have submitted for construction permits once his contract is signed. Since the permitting process is variable, he may create a schedule with assumed start and finish dates. Even if a firm construction start date is not known, the architect should insist that the contractor produce a preliminary schedule for review while the permits are pending. The schedule can always be updated later.

Once the agreement and all attachments are in order, and the insurance certificates, the schedule of values, and the schedule documents are received, the project can start. The architect will issue a document called a *notice to proceed*, which gives the contractor the right to occupy the site, commence work, and begin the construction duration period. Since this notice starts the contractor's clock for construction, the architect will usually coordinate the date of the notice with the contractor and the owner. Contractors may require some period to arrange for temporary utilities and field trailers, and to organize early subs to begin work. So there is usually a gap between final approval of the contract and actual construction start-up. There may also be a delay as well in awaiting permit approval. Since the amount of contract time available to build is critical to the contractor and the owner, coordinating the date of the notice to proceed is more than a common courtesy—it is essential.

In mobilizing on the site, the contractor may have a number of questions which the architect may be required to answer, or seek responses from the owner. Although the details of site mobilization are the clear domain of the contractor, these first construction interactions between contractor and owner are an area where the architect can act as an intermediary to help launch the project smoothly. Some areas where the architect may serve this role are:

- Obtaining owner information necessary for establishment of temporary utilities
- Assisting with determining site areas for setting up the field office, material storage, and recycling or trash collection
- Assisting with determining areas for parking for employees and contractor personnel, as well as site traffic routes.

In assisting the contractor at construction startup, the architect is performing a limited role as the owner's representative. The architect should remain aware that the means and methods of construction are the sole responsibility of the contractor. He should take care to limit his role to that of coordinator or facilitator, and always remind himself and the contractor that the final responsibility for construction operations on the site is that of the contractor.

PRE-CONSTRUCTION CONFERENCE

Near the beginning of construction, the architect should organize a pre-construction conference at the site. The pre-construction conference is an ideal time to set up lines of communication, discuss project documentation and meeting procedures, review submittal and project schedules, and discuss payment application and payment timelines. Here is an all-encompassing agenda for a pre-construction conference:

1. Contact information for key project personnel.
2. Identification of lines of communication for owner, architect, and contractor.
3. Review of construction schedule, including key deadlines and milestones.
4. Total contract value, retainage, and liquidated damages.
5. Schedule of values.
6. Permits or other regulatory issues affecting the project.

7. List of subcontractors.

8. Billing and payment timelines and procedures.

9. Change order documentation and request procedures.

10. Contractor job site office information and location.

11. Job site drawings & specifications.

12. As-built drawings.

13. Shop drawings and sample schedule.

14. Establish bi-weekly construction meetings.

15. Field observations by architects and consultants.

16. Quality control testing.

17. Field engineering.

18. Job site safety.

19. Hazardous materials.

20. Job site circulation and working hours.

21. Job site material storage (inside and outside).

22. Job site and/or building security.

23. Temporary or existing utilities

24. Coordination of owner-furnished material and equipment.

25. Coordination with other contractors.

26. Final inspection and punch lists.

27. RFI, submittal, and change order tracking.

CONSTRUCTION SCHEDULE

The contractor is responsible for preparing a construction schedule of how he intends to perform the work. This schedule is important because: 1) It sets a regular series of milestones against which the architect and contractor will measure progress, and 2) It is useful in assessing the rough percentage of work complete for line items on the schedule of values. If the architect had been keeping a project schedule for the owner prior to the contractor coming on board, his schedule is now subordinate to the contractor's. All progress during construction will be charted on the contractor's schedule. Depending on the size and nature of the project, he may not need to update it at each project meeting, but it should certainly be the centerpiece of a discussion on schedule conformance at each job meeting (see the Risk Hazard Flags).

When the contractor falls behind schedule, the architect should discuss the causes of the problem with him and how the schedule will be restored. If it is not possible to regain the lost time in the schedule, the architect, contractor, and owner need to discuss the consequences. If the contractor is behind due either to his or his sub's poor performance, the owner may agree to extend the construction time, but only with the agreement that he will not be subject to additional general conditions expenses. If the owner cannot extend the schedule without suffering damages, he may require the contractor to recoup the lost time at his own expense through overtime work or increased staffing. Conversely, where the schedule was affected by events outside the contractor's control, such as permitting delays, adverse weather, labor stoppages, police action, or other similar events, the owner may agree to pay the contractor additional monies to regain the lost time and restore the schedule.

 **Risk Hazard Flags**

Schedule Concerns

- Very little site activity visible
- Few subcontractors at work
- Field superintendent not regularly on-site
- Slow submittal and shop drawing submissions
- Contractor documents weather delays
- Contractor seeks excessive additional time for each change order
- Contractor has no plan to correct schedule lapses
- Contractor claims delay based on slow submittal or RFI responses
- Contractor does not produce updated schedule for each project meeting

If the owner agrees to a schedule extension for the contractor, he must prepare a revised schedule to update all work line items moving forward, so both he and the architect can track it appropriately.

PAYMENT ISSUES

When subcontractors or suppliers contact an architect with complaints that they are not being paid, it is a serious indication of trouble on a project. A skillful general contractor usually manages payment issues so that the architect never hears of them. Subcontractors rely on contractors for regular business and are reluctant to complain outside the family. When a subcontractor reaches out to the architect or owner for assistance, he is usually waiting on payments that are 60 days or more behind, and has become convinced the contractor is unreliable. It is also a warning shot to the owner that a lien against his property will be forthcoming if the situation does not improve.

An architect must be careful in these situations. A contractor may be holding monies from a subcontractor or supplier for good reasons, including poor workmanship, missing materials, inadequate job staffing, or poor schedule compliance. Subcontractors, like contractors, are also guilty of invoicing ahead of time for work they have not yet completed, or for invoicing more heavily for portions of the work that are not as valuable as his invoice claims. Architects who take actions that interfere with the contractor's legal right to manage his business can be accused of contractual interference in the agreement between a contractor and his subs. The architect should advise the owner to use restraint in such situations to avoid interfering with the relationship.

The only agreement that truly matters to the architect is the one between the owner and contractor, and that is the only one he need concern himself with. That agreement typically states that a contractor will pay his subcontractors and suppliers in coordination with his payments from the owner. In fact, the contractor signs an affidavit to this effect each time he completes an application for payment to the owner (assuming it follows the AIA model). Having been notified of a potential problem with payments to a subcontractor, the architect should:

1. Consult with the owner, who may wish to refer the matter to his attorney or offer advice to the architect on how to handle it, and/or,

2. Notify the contractor that a complaint has been received from one of his subcontractors and request an explanation of how much is owed and why it has not been paid.

3. If the contractor does not respond, or if his response is unsatisfactory, the architect should advise the owner to seek advice from his attorney.

With the attorney's guidance, the owner may begin requiring detailed periodic affidavits with each application for payment from the contractor, showing a list of payments and amounts owed to each subcontractor and supplier. In extreme cases, the owner may insist that joint checks be issued to the contractor and supplier to ensure that they are being paid as stated by the contractor. The architect will necessarily be involved in reviewing any action the owner requires of the contractor in alleviating this threat. As much as possible, he should attempt to work directly under the guidance of the owner's attorney and seek clear information on how he should handle future requests for payments.

Drastic action is necessary in situations where contractors are not meeting their payment obligations because such failures seriously threaten the ability of the contractor to complete the project in a timely fashion, and may impede the owner's right to use it for its intended purpose even if it is completed. If a subcontractor or supplier is sufficiently alarmed that he will not be paid, he may file a lien against the project. See Chapter 8 for a discussion of liens and strategies for handling them.

DEALING WITH PEOPLE

Effective Communication

There is a basic maxim of law that should also apply to architectural construction administration: If it is not written down it did not happen. The most fundamental rule an architect can follow in pursuing his work is to record key discussions and decisions. If this seems excessively formal to some parties in the business relationship, so be it. Memories and recollections of events and conversations are notoriously weak. Even moments after a conversation to discuss a difficult issue, parties to the discussion may walk away with drastically different opinions as to how the situation was resolved. Attorneys call the process of recording events *memorializing*. Design professionals may simply call it taking notes, keeping minutes or logs, making reports, or writing memoranda. Whatever a professional chooses to call the continual and diligent act of recording on paper or electronically, it will be his best tool and defense in negotiation, mediation, arbitration, or litigation.

In problem resolution, the best policy is to clearly state the problem without assessing blame and to seek proposed solutions from all parties. It is helpful during these sessions to have the subcontractor or supplier in the room if they are involved in the issue.

A discussion that remains fluid is best. That is, a discussion in which a moderator or leader (often the architect) keeps the discussion moving toward a solution without allowing it to deteriorate into a round-robin of blame. A discussion that focuses on problem-solving, while openly acknowledging that blame can be shared among several parties, is usually a more productive course than attempting to identify a principal culprit who must bear the bulk of the burden. Shared pain is a time-honored tradition in construction problem-solving.

Phone calls to the architect from the field are a necessary aspect of construction in any project. But they should not become the norm, because when they do this indicates that the field superintendent or his subs are not spending time looking ahead and planning the work. Each phone call should be followed up with a field instruction or RFI response from the architect. It is appropriate to ask the field superintendent or project manager to generate an RFI for any phone request he makes, even if the architect is able to respond immediately to the request. This is for the benefit of both parties.

E-mail communication is either a benefit or detriment to construction communication, depending on how an architect views such things. An entire chapter could be written on the problems, pitfalls, dangers, and difficulties of using e-mail heavily for communication. Since e-mail is not going away, and its ease of use almost guarantees it will expand to become the primary form of communication between contractor and architect (if not there already), perhaps the best course is to cite a few rules for effective e-mail use:

- Write formal e-mails. Use complete sentences. Write clearly and succinctly.
- Use one e-mail per major topic.
- Use specific subject lines.
- Assume the e-mail will be read in court someday.
- Do not send blind copies.
- Avoid overly long e-mails. Attach formal memos for long subjects.
- Retain copies (electronic, paper, or both) of all e-mails.

Project meetings need to begin promptly, follow a regular agenda, and move briskly. The leader of the meeting (typically either the architect or construction project manager) needs to keep the meeting on task and hold parties accountable when promised responses or documents are not available. A task list is a useful addition to the meeting minutes to remind parties of commitments. Issues that require prolonged discussion need to be taken "off-line," and resolved at a special meeting for that purpose or through some other means. The meeting minutes need to fairly and openly convey the issues discussed and any agreements made to resolve them.

Contractors are often required as part of their contract with the owner to conduct project meetings and keep meeting minutes. Some architects prefer to document the meetings themselves, reasoning that the keeper of the minutes can slant information more favorably toward his side. This should not be an overriding concern. Even if the contractor keeps the minutes, the architect has every opportunity to correct (and even expand) the record if the contractor does not accurately record the discussion. Even a contractor who is actively trying to use the meeting minutes to portray the architect in a poorer light than the discussion justified will reform if the architect repeatedly corrects his minutes at each

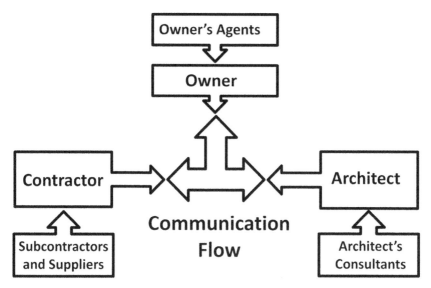

FIGURE 5-1
The communication flow.

project meeting. Whoever keeps the minutes should allow a reasonable time for meeting attendees to send him corrections, clarifications, or additions to the record, and diligently append whatever he receives. (See Figure 5-1 for an example of efficient project communication).

Adversarial Relationships

Adversarial relationships can develop quickly in construction. The pressure of budget and schedule, the questioning of intent and clarity in the documents, details that are missing or do not work, substitutions, the review authority of the architect, or dozens of other issues common to any project can flare into disagreements. Honest, specific expressions of disagreement are unavoidable, and are actually helpful in resolving construction disputes at an early stage. They focus the discussion and let each party express why he feels his position is correct. When these expressions do not lead to a constructive discussion, or degrade into personal attacks or expressions of distrust, they become destructive to the teamwork necessary for a successful resolution. It is easier to resolve problems in a professional environment. Be wary of a work site culture of intolerance. Abusive language and crude behavior is unprofessional, unproductive, and unacceptable. Construction is a profession with high ethical standards. Behavior that betrays those standards should not be tolerated, and will always make constructive problem-solving more difficult. See Figure 5-2 for simple tips for dealing with people in a professional manner.

Following are a few tips for engaging in constructive disagreement resolution:

- ***It's only business.*** Address other parties professionally. Never question their motives or integrity. This is especially difficult if the other party does not feel or act likewise. If the other party in a dispute is addressing the architect in an unprofessional manner, he should dampen emotions with statements such as: "I prefer to have a professional discussion with you," or "I think we should discuss this later when we can do so in a businesslike manner."

- ***Acknowledge their right to disagree.*** Usually, there is a basis for any disagreement. Something is unclear, was implied, or was understood incorrectly and triggered the dispute. Acknowledging this fact does not undercut the architect's argument. A statement such as: "I understand why you feel that way," or "I can see why you are upset by this," goes a long way toward defusing hard feelings.

- ***Review the evidence.*** Usually some piece of the construction documents is at the heart of the disagreement. It can often be worthwhile, particularly where the disagreement involves a subcontractor, to step back and look at the entire set of documents. For example: "I agree that the drinking fountain was not shown on the plumbing plan, but it was listed in the plumbing schedule, electrical schedule, and shown on the architectural plans and interior elevations." This *preponderance of evidence* technique ("we showed it

- Team Problem-Solving

- Professionalism

- Considerate
  Personal Relationships

- Respectful Conflict
  Management

FIGURE 5-2
Dealing with people.

four times and missed it once") is often effective in persuading parties focused on one aspect of the project that there is a larger picture to consider.

- ***Look for some middle ground.*** While an architect has to take care in negotiating on behalf of the owner (and must make clear that any decision is subject to approval by the owner), he is best qualified to work out solutions with the contractor. The architect is the authority on the construction documents; he is the one person with the most knowledge of what they contain and the unique knowledge of their *design intent*. Where disagreements threaten to disrupt the project, the architect can propose solutions to allow two or three parties to "share the pain" of a resolution to the problem. If an electric drinking fountain was shown on the architectural drawings, but not mentioned in the specifications or shown on the electrical or plumbing drawings, the two subcontractors have a valid complaint that they had no idea it existed. The general contractor could have recognized the problem and addressed it during bidding, but that is asking a great deal of a contractor during the short time he has to bid a project. Certainly, plumbing and electrical subs share some responsibility for reviewing the entire set of documents to look for such discrepancies and bring them to the architect's attention.

In situations where the drawings clearly do not indicate the scope sufficiently for a reasonable contractor to know it was in the project, the architect must explain this fact to the owner. Owner's reactions may range from unhappy acceptance of a change order to insisting that they are entitled to the missed feature at no additional cost. Where there is sufficient information on the drawings that the contractor or his subs should have been aware of the inclusion of the scope, the architect can argue that the contractor and his subcontractors (and perhaps the architect as well) have a shared obligation to provide it to the owner. In the case of the drinking fountain, if the plumbing and electrical subcontractors will contribute the labor to install the fountain, the general contractor may purchase it since it is arguably in his contract. If he declines, the architect can attempt to persuade the owner to purchase the equipment. Persuading an owner to pay a change order for equipment when the labor is provided at no additional cost is an easier—though not a sure—sell. The owner may ask the architect to participate in this shared arrangement with the argument that his drinking fountain is more costly as a change order than it would have been if included in the bid documents.

FIELD REPRESENTATION

Field representation has always been a bit of a conundrum for architects—particularly emerging professionals. There is a natural eagerness to see the work in progress, to witness the building rising from paper into reality. On the other hand, there is an equally natural wariness of becoming enmeshed in the means and methods of construction, or becoming the unwitting victim of a contractor's schemes for change orders. In between the idealism and unnatural fear is the reality of an architect's obligations to his client, himself, and his profession. An architect has three fundamental reasons compelling him to visit the job site during construction:

1. To review periodically the progress of the work (required by most owner-architect agreements).

2. To provide knowledgeable interpretations of the contract documents when requested to do so (required by most owner-architect agreements).

3. To reasonably satisfy himself that the owner is receiving the quality and full scale of work required by the construction documents (desirable from both a professional liability and a professional ethics standpoint).

An architect should take the role of field representative seriously, preparing for project meetings and bringing his talents to the table as the party most knowledgeable about the details of the construction documents and the needs of the owner. An architect can also act as an impartial observer of the contractor's performance, offering advice regarding schedule, quality, and scope concerns. Architects also bring superb problem-solving skills to any situation. Of all the professionals involved in the project, they possess the range of abilities to help the owner and contractor resolve hidden conditions, phasing, weather-related delays, or any of a host of conditions that present themselves during the course of construction.

Some owners, particularly developers and clients with in-house construction staff, have come to view the architect's construction administration services as an unnecessary expense. While they may recognize the necessity of having the architect on call to interpret the documents, explain unclear details, or check structural or MEP submittals, they otherwise believe the amount spent on construction administration services is not money well spent. This is unfortunate, as it fails to recognize the real value an architect brings to the owner in problem-solving, and in helping to ensure he receives the true value owed to him under the construction documents. The author of the documents, the architect, is the one professional ideally suited through training and experience to provide the owner with this benefit. The owner, simply put, suffers any time the architect is isolated from the act of construction.

Unfortunately, another fundamental issue also arises from this isolation. Problems that might otherwise have been observed by the architect at his weekly or bi-weekly site visits or project meetings go unnoticed until much later, giving the architect little leeway in mitigating their effects or working on less costly ways of resolving them. In cases of intermittent or reduced construction administration services, these problems are presented to the architect for resolution on short notice, or even after the fact. In these instances, the architect may be viewed less as a problem-solver than as the author of the problem, and an unwilling financial participant in its solution. The old adage of gossip applies to construction problems as well: "If you aren't in the room, they're gonna talk about you." Architects who are not allowed to participate in the early stages of a problem's growth are not able to defend their documents and intent, and therefore will invariably suffer more of the blame from the owner and contractor. For this reason, it is wise for the architect who is aware of a problem developing on the job site related to his documents to become involved—regardless of whether he is compensated or not. It is in his best interest to use his skill and talents to improve the situation for both himself and his client. Being part of the solution to a problem is not the same as admitting guilt as the cause of the problem. Architects must use great care in suggesting or conceding that their documents contained errors and omissions, or were otherwise responsible for the problem.

ELECTRONIC DOCUMENTS

Electronic documents used in construction come in several forms. The most common are portable document format (PDF) files, widely used to convert construction documents to relatively small files that can be shared via e-mail and easily printed in a variety of sizes. PDF files also possess the advantage that they cannot be edited. Construction documents are created using computer-aided drafting (CAD) programs, commonly saved with the *.dwg* suffix and readable by AutoDesk® programs, including their popular AutoCAD® program.

Architects routinely share PDF files with contractors and owners alike. They view a request for them to release electronic CAD files with concern, however. Unlike the limited usefulness of PDF documents, CAD files in the hands of the owner or contractor can be used for any number of purposes beyond their intended use. Architects worry, particularly in the case of some clients, that their electronic documents will be used in some nefarious purpose either to create additional liability for them or to deny them future fees because the client can use a single-purpose set of documents repeatedly or as the basis for a modified version of the architect's original work.

There is some validity, if somewhat overstated, to both of these worries. Because contractors and owners can have valid reasons for needing the electronic documents, they can claim to be entitled to them. For example, contractors and their subcontractors use CAD files as a base for coordination drawings, shop drawings, and as-built documents. Owners note that CAD files are useful to them as record drawings for maintenance or alteration purposes in the future, and allow them to overlay layers of information that can help them with internal relocations and operations. All of these uses benefit the project and the owner, so how can they be accomplished in a way that protects the architect against additional exposure from future illicit use?

The first way is to determine what each user needs. Contractors do not need (and probably do not want) an entire set of CAD files. A fire protection subcontractor, for instance, only wants floor plan, building section, and reflected ceiling plan files. An HVAC contractor may request only a building section, mechanical plan, and reflected ceiling plan. An owner may want the entire set of CAD files, but when pressed will probably admit that he really only needs CAD floor plans, reflected ceiling plans, façade elevations, and a building section. So the first step for the architect in protecting his interests in releasing electronic documents is to limit what he releases.

The second means of protection is to exclude the title block file from the transmitted CAD files, and strip out any associated references files or path directions that refer to them. The presumption

 *Risk Reduction Tools*

Electronic Document Disclaimer

Acceptance of these electronic documents implies agreement with the following conditions:

- These electronic files are not as-built documents and are provided to assist with the preparation of shop drawings and coordination of the project only.
- No express or implied warranties are made with respect to information contained in these documents.
- The architect has no obligation or liability for the accuracy of the information contained in these documents.
- The architect shall not be held liable for any consequential or special damages, or for any loss or profits sustained by user in connection with or arising out of use of these documents.
- The user is responsible to verify all as-built conditions in the field and coordinate their services and/or work as necessary.
- This information may only be used for shop drawing and coordination work for the original site and project for which it was prepared. This information shall not be used with other projects, and shall not be transferred to any other party for any reason.
- Reuse or reproduction of this media, in whole or in part, for any other purpose than that for which it was intended is prohibited.
- Revisions to the construction documents may be issued following release of this information. The architect is not obligated to provide an updated electronic form of this document.

is that any user will be responsible for using the CAD file as a new original, and therefore responsible for all information contained on the drawing. Though the author of the original documents, the architect should not place himself in position to accept liability for any future use or modification of the drawings.

Yet another method of limiting liability from the release of electronic documents is through the use of a standard disclaimer that places limits on both how the document can be used and the architect's responsibility for any consequences arising out of that use. To alleviate this worry, the architect can use an indemnity agreement to protect himself against claims based on uses of the drawings for other than their intended purpose. See the Risk Reduction Tools for a sample electronic document disclaimer form.

JOB SITE SAFETY

The architect is not contractually responsible for site safety. The general conditions of construction of most contracts require the contractor to be responsible for maintaining a safe work site at all times. Typically, this is accomplished through mandating compliance with the latest edition of the *Manual of Safe Construction*© published by the Association of General Contractors. Despite the fact that the contractor retains full responsibility for construction safety, this does not mean that any unsafe conditions can be ignored by the architect. An unsafe condition is one that the average architect or construction personnel would recognize as hazardous (see the Risk Hazard Flags).

 Risk Hazard Flags

Site Safety Liability Warning Signs
- Messy or disorganized job site
- Site accidents or OSHA citations
- No contractor safety program
- No weekly contractor safety meetings
- No safety review in project meetings
- No mention by architect of observed safety problems
- No response from contractor to noted safety issues
- Workers regularly failing to use fall protection or safety gear

Situations that may be considered unsafe are any in which there is the potential for danger or physical harm.

Examples: *Concentrated loads on a structure, work in trenches with no shoring, open flames in unsafe areas, open or unsafe areas with no barricade, and workers on scaffolding or roofs without harnesses or fall protection, and the absence of hard hats or safety glasses in areas where they are required.*

Because architects visit the site at infrequent intervals, they are not regular and consistent observers of the means and methods of construction. When they do visit the site, however, the architect is responsible for noting to the contractor any unsafe conditions he observes and demanding correction of them.

Whether or not the architect contracts for construction administration responsibilities, he must nevertheless be mindful of the importance of job site safety. Architects often cite the mantra that "the contractor is responsible for safety on the construction site." This is certainly true, but it does not relieve the architect or his consultants from the responsibility of bringing to the contractor's and owner's attention any activities that he considers unsafe or a hazard to others. The architect's concerns for safety are always present for four reasons: public safety, his client's property, the lives and health of the workers, and the possibility of personal liability (i.e., financial responsibility).

The architect's potential for liability as a result of responsibility for job site conditions will vary from state to state, and depends in large measure on that state's legal view of the role of the architect during the construction project. If the state where the project occurs sees the architect as little more than a construction manager with a role to play in "supervising" the activities of the contractor, then it is far more likely that the courts of that state will hold the architect responsible in the same manner it would the contractor, the construction manager, or even, under certain circumstances, the owner of the project. Where, however, the courts assign to the architect the more appropriate role during construction of administrator of the construction contract, with little ability or expectation of controlling day-to-day operation of the contractor, then the architect's liability is typically limited to deviations from professional care and practice, with virtually no responsibility for the contractor's means and methods, including his safety programs, procedures, training, and site conditions.

Sometimes the design professional's responsibility is a function of what he observes on the job site in terms of possible dangerous conditions and his ability under the construction contract to take curative action. The law in this area continues to evolve but is most frequently represented by the analysis employed by the New Jersey Supreme Court in the case *Carvalho versus Toll Bros. and Developers, et al.*, a 1996 New Jersey lawsuit. In that case Mr. Carvalho was fatally injured while excavating a trench during storm drain construction. The trench collapsed because the contractor failed to utilize a trenching box to shore the sides of the trench against liquefaction. The project engineer was named as a defendant but was dismissed from the case based on his contract for professional services, which relieved the engineer from responsibility for job site safety. The construction contract's front-end provisions placed all safety responsibility in the hands of the contractor, who alone was responsible for the means and methods of construction.

On appeal, however, the court determined that, while contractual provisions of this type were common, they only applied to protect the engineer from the contract claims of the parties to the construction contract and did not immunize the engineer from the claims of individual workmen who had no contract with the engineer. This was especially true in this case, because the engineer's field representative had observed the admittedly unsafe condition represented by the absence of the shoring device before the accident but had failed to advise the owner or contractor of the need to cure the dangerous condition. The engineer's potential liability was not only based on his duty to call out obviously dangerous conditions of which he was aware, but also on the engineer owing this duty to the owner in light of the impact on the schedule and progress of the work that resulted from safety-related accidents on the work site.

This case has recently been codified in New Jersey statutes and is often seen as the modern view of the contract administrator's responsibilities to the workers on site when it is foreseeable that an observed dangerous condition may cause serious injury or property loss, and when it is within the means of the professional to call out the problem for remediation by the contractor. The design professional should review with his attorney, professional liability insurance broker, or claims specialist the standard of responsibility observed in the states in which he practices.

Regardless of the state of practice or the state of the law, all design professionals can take some immediate common-sense measures to reduce the risk of ultimate responsibility for job site safety problems.

1. Make sure your professional service contract as well as the construction contract recites the absence of responsibility on the part of the architect for the contractor's work, including the means and methods of construction and the responsibility for the safety of the site or of the workers. The contract should include not only these standard recitations but also specific indemnity provisions in the event any agent, employee, or representative of the owner, the contractors, the architect or its consultants sustains any personal injury or property damage arising out of the work of the contractor, whether or not the injury or damage is the fault of the contractor (specific language must be guided by the laws of the jurisdiction where the construction occurs).

2. The architect should insist that the owner also include in any construction contract the full range of insurances (general liability, auto, builders' risk) and that the contractors and all of their subcontractors name the owner and the architect and its consultants as additional insured parties under policies of commercial general liability insurance. These policies should be considered primary and without regard to the coverage maintained by the additional insured parties.

3. The owner should be advised not to permit the contractor to commence construction operations on site until the appropriate certificates of insurance have been obtained and verified. The certificate's coverage periods should be checked and the owner must be given the opportunity to replace the policies should the contractor allow the policies to expire, with a back charge to the contractor for these expenses.

4. Where the contractor is responsible for submitting a safety plan, this portion of the contract should be enforced. The owner may ultimately be responsible administratively to the Occupational Safety and Health Administration (OSHA) if he regularly turns a blind eye to the contractor's faulty safety operations that result in serious injury or death. While current trends in the law continue to hold the line against OSHA, finding responsibility against design professionals who are not functionally "in charge" of the work site, OSHA continues to push the boundaries on this issue where a designer's faulty plans and specifications are a possible cause of work site injuries.

5. Avoid the "see no evil" approach to construction administration. If the architect, his field representatives, or consultants see or learn about dangerous conditions or activities, these must be reported immediately in writing to the responsible contractor, with copies of the notice to the owner and construction manager. The immediacy or magnitude of the danger may require even more direct action. Pick up the telephone and give immediate notice of the condition and follow up by memorializing the notice. Using the fax can solve both issues at once. It is important to remember that such notice does not obligate the architect to dictate the proper method of remediation of the condition. After all, the contractor, not the architect, is the expert in means, methods, and safety. There is of course a fine but definite line between warning the contractor of a dangerous situation and interfering by directing his activities. Follow-up can be limited by obtaining assurance from the contractor that he is aware of the problem and his brief description of how he intends to cure it. This information, if not provided by the contractor in writing, should be memorialized by the architect and faxed to the offending contractor.

6. During project meetings the architect should maintain a regular agenda item to reconfirm that weekly safety meetings are occurring, and note for the record any job site safety incidents or injuries that have occurred, and how the contractor will be avoid them in the future.

CONSTRUCTION TESTING AND CERTIFICATION

An architect's specifications call for a wide variety of manufacturer's certifications and field testing of materials, depending on the nature of the project and how critical the individual product is to meeting the owner's needs. Certifications are often required for framing materials such as lumber, light-gauge and cold-formed framing, drywall, and structural steel. Field testing and certifications can be called for in areas such as:

- Design of concrete mixes and preliminary tests
- Identification and test data on reinforcing steel
- Identification and test data on structural steel
- Field testing of mechanical piping and equipment
- Field testing of electrical wiring and equipment
- Testing and balancing of HVAC equipment
- Furnishing samples for testing
- Determining that material certificates for specific products demonstrate conformance with the specification requirements
- Delivered concrete
- Flatness of concrete surfaces
- Base compaction
- Bearing capacity of concrete or asphalt
- Paint or surfacing thickness.

When test results come back below the required criteria for the project, the architect must take the lead in determining what, if any, action needs to be taken to correct the problem and whether a change in specifications is warranted. This situation most often occurs with concrete mixes, where initial seven-day breaks of test cylinders taken the day of the pour may show that the predicted ultimate strength of the concrete will fall below the design strength. Typically in these situations, the architect will convene a discussion among himself, the testing agency, the structural engineer, and the contractor to determine how to interpret and respond to the results.

If, for instance, a second cylinder from the same pour shows different results, the cylinder in question may simply be an anomaly, an odd outcome resulting from any number of reasons. The team may decide to break a second cylinder at 14 days to see if the predicted strength improves. Seven-day breaks, while normally reliable indicators of ultimate strength, can under-predict strength if they are stored improperly or allowed to cure too quickly. In the event the team decides that the test results are valid, they may elect to perform a core test on the concrete in place to determine its actual strength. If the contractor objects to paying for this test on the basis that he feels the concrete test method is flawed and he has met the contract requirements, the owner may offer the compromise that he will pay for core testing if the concrete meets the design strength, but the contractor will pay for testing if the concrete falls below strength.

The point is that field test results of any kind require a team approach to assess and resolve. The contractor should be included in all discussions, as well as concerned subcontractors or suppliers. It is important to remember that single test results are not always valid, and even when they are, slightly compromised installations may still be determined to be more beneficial to the owner than their removal for subsequent reinstallations that are fully-compliant.

LENDER AND OTHER INSPECTORS

Lenders may or may not accept the architect's certification of work completed on a project. Often they will engage an independent inspector to review the progress of the work associated with each application for payment. Similarly, the owner's insurance carrier may have an inspector visit the site to review fire protection, roofing, or structural elements. This is particularly the case with projects required to meet Factory Mutual® (FM) requirements. The architect has no formal responsibility to consult with any of these inspectors, but the owner often calls on him to:

- Provide construction documents for the inspector's use
- Coordinate visits with the progress of the work (a particular issue with applications for payments)

- Answer questions from the inspectors regarding compliance with their requirements.

MUNICIPAL CODE OFFICIALS

Code officials are usually much more attuned than architects to the details of installation, particularly in the mechanical, plumbing, and electrical areas, that are likely to result in threats to life safety. In most states, code officials attend a regular series of continuing education seminars to educate themselves on areas of construction that are resulting in fires, structural collapse, water damage, injury, and death across the nation. In that sense, they are truly more expert on the fundamental purpose of codes than are architects and engineers. Where architects are looking at codes more broadly, in terms of the major life safety issues such as maximum occupancy, egress capacity, and fire separation areas, code officials are most concerned with the margins.

Partly for this reason, municipal code officials are legendary for their ability to annoy and confound architects. It would be easy enough to chalk up the history of difficult relationships to competing egos and a desire of each to reign supreme as *master interpreter of the code*. In reality, though, the relationship is much more complex. Architects and code officials largely agree on the big issues of a project (there are exceptions, of course), with the arguments occurring over small details mostly considered arcane by the architect. Following are some of the common areas of disagreement between architects and code officials.

Completeness of Plan Reviews

Architects expect that code officials who issue a building permit have performed a complete and thorough review of the plans and marked any item that is not in conformance with the code. Invariably, a code official performing a field inspection will flag an item that should have been caught in permit review but was not. These flagged items are usually small (door swings, hardware, or the like), but they embarrass architects, generate change orders, and annoy owners.

Certificate of Occupancy Inspections

Late in the project, with the contractor pushing to turn the building over to the owner, the municipal inspectors stream in to perform certificate of occupancy inspections—their last review before the facility is opened to the public. Among these inspectors is the fire code official (or fire marshal), who wields enormous say as the final decider of life safety concerns. These inspectors can sometimes cause tremendous upheaval (and change orders) with demands late in the project for: 1) Additional smoke or heat detectors, 2) Additional emergency lights or exit signs, 3) Additional or altered locations for fire extinguishers, and 4) Additional fire protection or life safety signage. Their motives are pure, no doubt, but flagging critical items at such a late date sends the contractor and his subs scurrying to install the necessary items and baffles the owner as to why critical life safety items were missed by his architect. Some of these last-minute emergencies can be avoided if the architect and contractor work together to schedule an informal walk-through with the fire marshal or building code official in advance of the more formal certificate of occupancy inspections. At the informal walk-through, the architect can assist the contractor by both raising issues such as those cited above, and by discussing with the inspectors the rationale behind some items. The architect's presence and a less formal tour of the facility may help to smooth the path for the later inspections.

6

Under Construction—
Managing Documents

Managing the paperwork is a necessary requirement of almost any business endeavor. The difference in construction is that each document a design professional handles has potential liability attached to it. Every document, whether a simple request for information or a standard submittal, carries with it a risk of error, misinterpretation, or misjudgment. Combine this hazard with the sheer volume of paper flowing across a busy architect's or engineer's desk and the speed at which construction responses are required, and it isn't hard to see why even seasoned construction administration professionals can become overwhelmed by this responsibility—and the consequences of not meeting it fully.

During construction, design professionals have dual obligations to their client. These two roles can loosely be termed *interpreter* and *verifier.* As the creator of the construction documents, the architect is called upon to interpret them when instances arise where his intent is not entirely clear from the information shown on the plans and in the specifications. Secondly, the architect is part of the verification process, confirming that the specified products are being provided to the project, of if they are not available, that products of equivalent quality are installed as replacements and that the owner is credited with the price difference. Another aspect of this verification role is to make periodic visits to the site to judge, in a general way, whether the work is being constructed in accordance with the documents. This second obligation is actually shared among contractor, subcontractor, and design professional, but as we will discuss later, the architect is sometimes cast in the role of "quality cop" in ensuring that less conscientious contractors are not free to take advantage of an unaware or absent owner. An architect was once heard observing: "We want to be a resource for the best work, but we end up being the building police for the worst."

Both of these obligations involve careful review of documents required as part of the contract between the owner and contractor. Certainly the most volume of work is generated during the submittal process, when a veritable flood of paper comes across the design professional's desk for every product noted in the specifications as requiring a submittal. Particularly at the onset of construction, submittals are time-sensitive documents. Perversely, those requiring the quickest responses often present the most potential liability and require the involvement of the architect's consultants and his own careful attention to coordination issues. Concrete and its reinforcement, masonry, cold-formed framing, structural steel, joists, curtain walls, and HVAC systems are only a few of the critical submittals that must be reviewed by architect and engineer--and usually under short time constraints.

The other documents associated with construction are less numerous, but potentially just as deadly. Requests for information (RFIs), requests for substitutions, applications for payment, and all the assorted documents surrounding the construction of a project and management of a contract require a design professional's review and response. They also require something just as important. They require management and knowledge of the contracted response times.

EFFECTIVE DOCUMENT MANAGEMENT

Contractors handle huge amounts of paper. Not only do they generate all the information that crosses the architect's and his client's desk, but they receive volumes of information from subcontractors and suppliers, and yes, from design professionals as well. Some handle this deluge better than others, but the best of them are truly masters of document management because they have overhead available for staffing a project, often employing a dedicated project manager, field superintendent, and administrator on larger projects; and they use specialized database software that enables them to log, track, and cross-reference documents during the course of the project.

Unless they are practicing in a large firm, architects and engineers typically do not have either of these advantages. A design professional often manages several projects at a time, and support staff in most professional offices is spread thinly, and they manage the marketing, invoicing, and receivables first and everything else second. Submittal and request for information logging and responses are often the sole responsibility of the project architect or engineer. Submittals on even mid-size projects can number in the hundreds, and they roll into the office fast and furious in the first few months of a project. Similarly, RFIs can accumulate rapidly as subs and suppliers dig into the documents (sometimes for the first time) after they have been awarded the work by the contractor.

 ### *Risk Hazard Flags*

Effective Document Management During Construction

- Log It
  Keep a dated log of incoming and outgoing Submittals and RFIs.
- Document Verbal Responses
  Document phone or verbal RFIs as an architectural clarification.
- Document Field Instructions
  Initial and date field changes on the contractor's field set and architect's office set.
- Use Sketch Reponses
  Issue sketch (SK) 8½x11 or 11x17 responses for limited drawing clarifications.
- Use Construction Bulletins
  Issue construction bulletins for more comprehensive changes or clarifications.
- Respond within Contract Time Limits
  Always respond within the time obligations of the owner/architect or owner/contractor agreements. Document the reason for any delays in doing so.

Although submittals are increasingly being submitted to the archi-tect in electronic form, the profession still requires handling a large amount of paper. To manage this information, whether in paper or electronic form, design professionals have developed some simple techniques. They share the common characteristics of organization and sequence (see the Risk Hazard Flags).

Techniques of Managing Documents

The following techniques help with managing CA documents:

- ***Create a submittal schedule:*** Work with the contractor to develop a mutually agreeable schedule for submitting and returning submittals, focusing on those that are "mission-critical" to maintaining the construction schedule.

- ***Log all submittals:*** This can be accomplished with a simple spreadsheet program. Record when they come in; when you send them back; and what you did with them *(approved/ approved as noted/revised & resubmit).*

- ***Log all RFIs and responses:*** This can be as simple as clip-ping a copy of the original RFI (whether in fax or e-mail form) into a notebook with the dated response attached to it. If the architectural practice is more electronic, keep PDF copies of originals and responses in electronic folders. Each one should be dated.

- ***Record all field instructions:*** When an architect makes field changes, he should mark them on the general contrac-tor's field set (initial and date) and on his office set as well. If it is a critical change, make sure it is recorded in a memo or e-mail to the contractor and client.

- ***Document phone conversations:*** It is often a good thing when a superintendent has developed a relationship with a design professional in which both feel confident in discussing issues over the phone and resolving them verbally. This can be a quick and efficient method of dealing with small issues, and deciding them in a timely manner to keep them from metastasizing into larger problems. If the architect is comfortable with such a relationship, he should maintain it through the course of the project. But in doing so, he must take care to document the phone conversations in a quick e-mail or fax (pick one method and stick with it), and copy the project manager so he is clued into the resolution. Logging phone conversations and a short summary on a spreadsheet is also a useful technique.

- *Issue sketch responses:* All changes do not have to be full-size, formal construction bulletins. Issue sketch responses quickly to clear up issues that need immediate attention and don't represent scope changes as much as clarifications. Issue SK series sketch responses for architectural issues; SSK for structural issues; and MEPSK for mechanical, electrical, and plumbing issues. Always issue sketches in numerical order, and always date them.

- *Issue construction bulletins:* When substantial changes occur to the documents, perhaps including a number of sheets and more than one discipline, issue a formal construction bulletin with a summary explaining the changes. Always clearly cloud and mark changes to the documents.

Timeliness

One issue that constantly bedevils design professionals is the feeling of being pressured to respond quickly to field questions, problems, and discrepancies in the documents. By nature, architects and engineers like to study a problem and spend the time necessary to come up with the most economical and elegant solution. This is desirable, of course, but one factor that often seems to work against this is time. Time lost in construction has real value, to the point where contractors will take the risk of implementing their own solution to a problem instead of incurring a delay or additional expenses in waiting on a design professional to render his decision on the best course. Architects should always keep the value of time in mind when responding to RFIs or construction problems. Judgment is important, but also is remembering that the ideal solution delivered late may end up costing more than a timely decision, ideal or not. Conversely, however, this time pressure alone is a good reason to resist efforts by the owner to require an inflexible time limit on responses to the contractor. In the architect/owner agreement, or in the contract documents, the architect should insist on a reasonable amount of time to respond to submittals and RFIs. The architect should argue for no contractual time limits because his responsible service to the owner requires an inquiry into the complexity and circumstances of the problem. Some problems arc casy; others are not. The architect deserves the professional right to spend the minimum time necessary to solve the problem in the best interest of the owner. At the very least, the architect must know the contractual obligations for response times on submittals, pay applications, and RFIs so he does not incur liability for delays.

REQUESTS FOR SUBSTITUTIONS

On practically every construction project contractors will ask to install products that are different from the ones specified by the design professional. This occurs for various reasons, ranging from the valid and practical (the specified product is no longer available) to the near-sinister (the contractor substitutes a less expensive product and makes more money). Most contractors, in fact, are not making the substitution request for themselves. They are passing up the line a request from one of their suppliers or subcontractors, who are likely making the request mostly for their own convenience, or because a specified product is not available. Bidders often assume that a product they prefer will be considered equivalent to a specified product. They prefer to use items they can purchase at a competitive advantage, that they are most familiar with (and can install efficiently), or that they genuinely believe to be the best product for the purpose. Some also propose substitutions because they are the exclusive service providers for the substituted product. These bidders know their competitors and their costs very well. They also know how their product compares to the architect's or engineer's specification, and their history of obtaining successful substitutions on other projects. To get the work they take the risk that the design professional will accept their product as equal to the one he specified. Often they find language in the specifications encouraging them to do so ("...or architect/engineer approved equal").

This is not necessarily bad for the client—or the architect. A sub-contractor earns his place in the project by being the lowest bidder in his trade. Enough low trade bidders add up to a winning overall bid, which, if it is low enough, equates to a happy client more likely to give his architect or engineer future work. The flip side, of course, is that substitutions can pose a number of risks for design professionals, including: coordination problems, color issues, added cost for other trades, schedule complications, and operational differences. Any one of these issues can result in poor product performance and unhappy clients. Substitutions could be considered annoying for any of these reasons, but the main reason most design professionals despise them is because they devour professional time like a hungry wolverine. Consider the following path of one substitution request:

- A submittal arrives for an *Illumax* light fixture. Problem: The architect specified *Glowmaster*.

- The architect rejects the *Illumax* fixture, pointing out the note in his specifications stating that "substitutions will only be considered when accompanied by a point-by-point comparison to the specified product," and that substitution requests cannot be made via submittals.

- A week later a more detailed submittal arrives. It is hardly a point-by-point comparison, but the architect digs into the details for a couple of hours and judges the fixture to be largely comparable. But there's one glaring question that has to be answered: Is it available in the right color?

- In another week a color chart arrives. The stock color is close to the one selected as part of the interior package, but not quite there. The architect knows client will want the brighter color. He rejects the *Illumax*, finally and irrevocably.

- The contractor calls the architect the next day. Because so much time has elapsed in reviewing the substitution request, it is now too late to order the *Glowmaster* and meet the completion schedule. What product can be had in time? Of course, the *Illumax*.

It's no wonder, then, that many design professionals pepper their plans and specifications with *"No substitutions allowed"* notes. Such notes are not permitted on public work, but the techniques in this book deal with privately-bid projects. Unfortunately, these notes do not work because the architect or engineer is not the final arbiter of what goes into a project. That person is the owner, and contractors know they can appeal to an owner's purse, sensibility, or ego to persuade him to accept a substitution if he is convinced it is in his best interest. For that reason alone, substitutions will be around as long as cockroaches. They are an ingrained part of the bid process and can survive a nuclear winter.

So if substitutions are here to stay, design professionals must manage them in a way that will protect the owner, minimize time consumption, and perhaps even save a little frustration along the way. Design professionals should use the following three-step process designed to give themselves more control over substitutions:

1. ***All Bidders must certify they are bidding to the plans and specifications.*** This step is necessary to ensure that all base bids are based on the same information. Smart contractors already require that their subs and suppliers state their bids are based on the plans and specs, so they will have no objection to this certification. As previously noted, though, this is unlikely to eradicate substitutions since they are embedded in the bid process as a way to gain a competitive advantage. However, inclusion of this note in the bid documents at least gives the owner and the architect a foothold in the event of a dispute. The next two steps represent more robust methods of avoiding substitution disputes:

2. ***Flush out substitutions during the bid process.*** In private bids, control substitutions by allowing alternative bids in your specifications (see Figure 6-1). This language allows the contractor to provide a second, lower bid to the owner by listing proposed substitutions on a separate bid form, along with the credit associated with each. The use of this procedure actually encourages the contractor and his subs to seek out and propose alternative products earlier during

Request for Product Substitution

Project Name: Project No.

To (Project
Architect):   FAX:

From (Contractor):   FAX:
                                                                Date:

Substitution Request: *(Include drawing and specification reference)*

Reason for Substitution Request:
(Attach documentation supporting why the product is equal to the one specified)

Contractor's Proposed CREDIT if the Product Substitution is accepted: $_____

Architect's Response to Contractor:

NOTES

1. By submitting this request, the Contractor certifies that he has thoroughly reviewed the Construction Documents, and a substitution for a product of equal or superior quality to the specified product is warranted for the reasons listed.

2. The Contractor must submit all Product Data necessary to demonstrate that the requested substitution is equal or superior to the specified product.

3. The Architect's Response to the Contractor's Request for Product Substitution does not constitute a directive to proceed with Changes to the Contract Scope of Work. The Contractor MUST seek express approval from the Owner if changes in the Contract Sum or Contract Time may result from executing the work according to the Architect's Response.

FIGURE 6-1
Sample request for product substitution form.

the bidding period. The advantage of using this process is that the owner and design professional are made aware of possible credits early, and the contractor gains the benefit of proposing them in the bid process. It promotes and provides some order to something that is going to happen anyway: contractors trying to gain a competitive advantage by using substitute products. It also provides more leverage to the owner if the contractor ignores this opportunity and later requests a substitution.

3. ***Control construction phase substitutions.*** The third and final step in managing substitutions is to provide a procedure for submitting them once the project is underway. Despite encouraging contractors and subs to put their substitutions on the table early, despite blanket statements that no substitutions will be considered after commencement of the project, and despite the architect's steely cold responses to the contractor whenever he mentions a substitution— they will still occur. Without encouraging them, the architect should create a procedure for the contractor to follow in submitting substitution requests. It should be a fairly rigorous procedure, as the contractor literally has the "burden of proof" for demonstrating the equivalence of the substituted product. Having provided a bid form for the initial opportunity to submit his substitution requests (of which the architect can frequently remind him), the architect should insist that his substitution procedure be followed. The architect's stance on substitutions should be consistent, and require that his procedure be followed even if the contractor persists in attempting to make a *substitution by submittal*, or claims that the issue is moot since the product has already been installed. The procedure can be described in a simple form in the project specifications (see the Risk Reduction Tools).

The product substitution form requires the contractor to make a declaration that the product is equal or superior to the specified product, asks him to provide backup data proving this fact, and requests a change order credit for the substitution. The implication (which many design professionals state explicitly) is that no substitution will be considered without an attached credit.

Risk Reduction Tools

Bid Alternate Sample Specification Language

The contractor shall include provisions for the following bid alternates as listed in the form of proposal, and shall state the amount, in dollars, to ADD TO or DEDUCT FROM the base bid contract sum. If there is no change in the amount of the base bid contract sum, then write "NO CHANGE" in the form of proposal. The contractor shall be bound by all requirements and conditions of the construction drawings and the project manual as they relate to the bid alternates.

1. Bid alternates quoted on the contractor's form of proposal will be reviewed and accepted or rejected at the owner's option. Accepted bid alternates will be identified in the owner/contractor construction agreement and included in the total contract sum.

2. The contractor shall coordinate all related work and modify all surrounding conditions as necessary to integrate the work of each accepted bid alternate.

The Alternative Bid Process

The owner is under no obligation to accept the alternative bid if he finds the proposed substitutions objectionable. Nor is he under an obligation to accept every proposed credit. The owner may select a low bidding contractor based on the alternative bid form and negotiate which credits to accept or reject based on his design professional's recommendation.

There are challenges to the alternative bid procedure.

Fairness Issues

Contractors may feel the process is less fair, since the alternative bid form basically invites a bidder to propose a lessening of quality in the project, and to gain a competitive advantage from doing so. What preserves the integrity of the project is that contractors are savvy enough to realize that owners don't want a wholesale reduction in the quality of their result. They simply want to take advantage of any equivalent products that can save them money. And contractors and their subs are more likely to be aware of these products than design professionals. Design professionals should also stress to bidders that the owner's first intent is to select the low bidder of the plans and specs. If those bids are within his budget, he would have little reason to look at the alternative bids.

Shopping Ideas

Contractors are suspicious that owners will take their substitution ideas and give them to the bidder they otherwise prefer. This is a valid concern, and places a burden on the architect/engineer to educate the owner on the fair and proper use of this procedure, and take pains to assure contractors that their bids will be private and independent.

Private Bids Only

This procedure cannot be used for public bids, where public bid laws in each state (or federal statutes) would not allow wholesale substitution proposals during the bid period.

Full Credit

Owners may express concerns that the credits offered on the alternative bid form are not full value. The response is that they probably are not. The contractor may make a judgment that he will retain some portion of the credit to bolster his profit, provide a cushion for negotiation with the owner, or simply to provide some contingency for a few of the above-mentioned problems that often result from substitutions.

Substitutions are an inherent part of the construction process. Using the procedures in this section will enable the architect to flush them out early, use them to his client's advantage, and better manage the late ones that will inevitably occur.

REQUESTS FOR INFORMATION

Requests for information (RFIs) are the means through which a contractor seeks additional information that is not present, or is unclear, in the construction documents. Most of the time RFIs are simple requests for clarification of mundane information that is necessary for completing the work, such as a missing framing dimension or a wall type that was not indicated on the drawings. Although the RFI should always come from the contractor, the questions represented in RFIs are usually generated by a subcontractor or supplier. In Chapter 4 we discussed bid requests for information. These occur during the bidding phase as contractors and their subs pore over the drawings and specifications to perform quantity takeoffs and prepare their bids. Responses to RFIs during the bidding phase usually are concerned with scope issues, or items mostly related to quantity and quality of materials. The responses are incorporated into the bid, and ultimately into the construction contract, and therefore do not represent a change order to the owner.

RFIs submitted during the construction of a project are quite another matter. Responses to RFIs that generate additional scope, or make a fundamental change in the contract, can result in a change order request to the owner. For this reason, architects and engineers must be particularly sensitive to RFIs containing language that hints at potential change orders. Some contractors may also use this process to ask for product substitutions, build a case for a delay claim, or seek means and methods shortcuts to save them time and money. Following are a few examples of these types of RFIs:

- *The door schedule calls for seven foot high doors but the elevations indicate eight foot high doors. Our bid was based on the door schedule.*

- *Cold-formed framing gauges are shown on the drawings, but the specifications require the contractor to provide engineering of the cold-formed. We assumed the drawing gauges were correct.*

- *The scale on the plumbing plans is shown as $^1/_8$ inch=1 foot, but all other plans in the set show a scale of $^3/_{16}$ inch=1 foot. Which is correct?*

Each of these examples indicates a subcontractor submitting an RFI for the purpose of positioning himself for a change order. In each case, he is already staking out a position that his bid was based on a portion of the construction documents, and any discrepancy that varies from that bid assumption is just cause for claiming more money. The architect's response to these types of RFIs needs to be carefully worded to accomplish three goals: 1) Provide the clarifying information in a timely manner, 2) Not to add any ammunition to a likely change order request, and 3) Use the RFI response to communicate to the contractor or subcontractor that he will defend his documents.

RFI Responses

When responding to an RFI, the architect should keep two basic ideas in mind. First and foremost, the architect should rely on the construction documents. Perfect or imperfect, any set of plans and specifications contains a great deal of information that obligates the contractor to investigate, to verify, or to confirm. The contractor will complain, perhaps loudly, that the architect is placing code compliance and engineering burdens on him that properly belong to the architectural and engineering team. If discrepancies exist in the construction documents, the contractor will also point to the section of the documents that best supports his case.

All RFI responses, even those centered on engineering questions, should flow through the architect. Cumbersome though it may be, the architect should relay RFIs to and from his consultants. He should be prepared to modify the engineer's language, if necessary, to alter or remove any verbiage that the contractor can use in seeking change orders or delay claims.

Use the Documents

Reference details, specifications sections, drawing conventions, notes, standards, and any other aspect of the construction documents that provides a reasonable argument that the work is the responsibility of the contractor.

Acknowledge Discrepancies

Construction documents are full of discrepancies. It does not necessarily diminish the architect's argument to acknowledge that two parts of his documents contain different information. In fact, he may turn this to his advantage by noting that the subcontractor or supplier is responsible for reviewing the project scope and seeking clarification to any discrepancies in the plans or specifications. Where two references contradict a third that the contractor is citing, the architect should note this as well. The fact that a contractor has ignored multiple instances of evidence to the contrary makes his claim less defensible.

Remember the Audience

The owner has a contract with a general contractor or construction manager. Even though the RFI may have been generated by a subcontractor or supplier, it came to the architect through the general contractor—and that is the party to whom he will be responding. The architect must keep in mind that he is responding to the general contractor, not to any other party. This is significant because the general contractor has more comprehensive responsibilities than his subcontractors. A general contractor is responsible for the entire set of construction documents, for plugging gaps, for general conditions, and for all elements of the contract for construction. Subcontractors may be able to view their scope of work narrowly, but general contractors cannot. In responding to an RFI that argues that one portion of the documents—say the plumbing drawings—does not indicate a drinking fountain that is clearly indicated elsewhere, the architect will reply that the general contractor is responsible for installing the drinking fountain since it appears in three other locations in the documents. The plumbing subcontractor may or may not have a legitimate cause for a change order with the general contractor, but that is not the architect's concern. The missing drinking fountain should, of course, have been shown on the plumbing drawings, but it was indicated elsewhere quite clearly and even appeared in the specifications. In the court of RFIs, the

 *Risk Hazard Flags*

Requests for Information (RFI) Tips

- Answer Narrowly
  - *Don't go beyond the question*
  - *Be careful not to expand scope*
  - *Focus on what to build; not how to build*
- Use References
  - *Reference the plans and specs in your responses*
- Answer Completely
  - *Don't leave the question open*
  - *Avoid prompting a second RFI with ambiguous answers*
- Answer in a Timely Manner

weight of evidence is on the architect's side that the general contractor should have known the drinking fountain was in his scope of work, and is therefore responsible for it under his contract with the owner (see the Risk Hazard Flags for more RFI response tips).

Means and Methods

The architect should tell the contractor what to build, but not how to build it. If the contractor's primary responsibility is means and methods (or how something is built), then the architect's turf is the depiction and articulation (pictures and words) of the design intent—in other words: what is to be built. Contractors sometimes express their frustration with the architect or his documents through RFIs when they do not clearly understand how to build something. They will insist, directly or indirectly, that the architect tell them how. In essence, they are inviting the architect to err and will pounce on him when he does. The prudent architect will refuse this bait. In responding to RFIs (even repetitive ones asking the same question), he should continually address the *what* of the documents by clarifying the design intent, while avoiding telling the contractor *how* to build the project.

Sometimes, particularly in the instance of *performance specifications*, the contractor is expressly required to seek assistance from a consultant—or even signed and sealed drawings from a licensed professional—who works for and is paid by the contractor. Other times, the need for engineering assistance may not be expressly stated in the documents but is clearly apparent from the nature of the work or the requirements of the project. However much the contractor begs, pleads, threatens, or cajoles, the architect must resist the temptation to tell the contractor how to build in an RFI response. It is a path to liability trouble.

SUBMITTALS

Submittals are the key component in the simple and effective quality assurance process of construction. As an architect's desk groans under the weight of a pile of submittal packages, he probably questions whether this process is simple—or even necessary for that matter. As a pure management tool, the submittal process is both. Submittals engage the triad of design professional, contractor, and subcontractor/supplier in an efficient, albeit paper-driven, system to ensure that the products the architect selected and specified, and the building systems drawn, are being provided to the project. In the pure submittal process, the subcontractor or supplier reads the drawings and specifications, prepares a submittal package based on what was specified, and submits it to the architect with a sufficient number of copies for the architect to review and approve. If details were missing in the specifications, a color needs to be selected, a finish chosen, or a mounting arrangement confirmed, these are marked for the architect's response. If a subcontractor needs additional dimensions, he may request these on a shop drawing. In the case of steel submittals, much of the detailing of the connections and anchorage is prepared by the steel detailer for review by the architect and his structural engineer. In other cases, such as fire protection systems, truss design, shoring design, or cold-formed framing, the project specifications may require the contractor or his subcontractor to engage a licensed engineer to document this work (referred to as a performance specification) in a submittal for the architect's review.

The subcontractor or supplier sends the submittal to the general contractor. He reviews it for completeness and coordination issues, and forwards it to the design professional, who performs his own review. If the submittal is the subject of engineering design, the architect will send it to the engineering consultant of the appropriate discipline (typically, mechanical, electrical, plumbing, or structural).

If the submittal is incomplete or incorrect, the design professional may reject it and ask for a resubmission. If only minor corrections are required, he may mark it "approved as noted."

When the submittal process works, it is very effective at flushing out coordination problems early in the construction process before they reach the field. When the submittal process breaks down, there is significant risk for the design professional. This can occur at numerous points, including:

- The subcontractor submittal is generic and general. The design professional must make assumptions or reject the submittal and risk delaying work.

- The general contractor does not review the submittals; he merely stamps them and passes them onto the design professional. As a result, potential coordination problems among construction trades are not caught. Weak submittals that should have been sent back to the sub for revision waste time going to and from the architect and/or engineer.

- The design professional is rushed and does not spend adequate time reviewing submittals. He misses conflicts or does not respond to subcontractor verification requests on the submittal, increasing the potential for field problems.

- The contractor fails to follow the submittal schedule he helped to create. Late submittals pile up, interfering with the orderly flow of submittals and placing a burden on the architect to review them quickly.

- Due to late submittals by the contractor or tardy reviews by the architect, both parties begin taking shortcuts in the process: submittals bypass the general contractor (sent from subs directly to the architect or his consultants, or submittals bypass the architect (sent directly from the general contractor to his consultants). This is a high liability practice. The architect is essentially accepting, blindly, the review of his consultants on his behalf. The architect should always insist on submittals being reviewed by the general contractor and submitted through his office (where of course, he reviews them).

The submittal process works well when all parties participate fully and professionally, and three sets of eyes are looking at each submittal. When one of the three participants is not doing his part, it removes one layer of review and places an extra burden on the other two. When either the contractor or the design professional is not carefully reviewing the submittals, there is an increased potential for uncaught problems to be relayed to the field, where their resolution is always more difficult and costly.

Checklist for Submittal Checking

The following is a suggested checklist for submittal checking:

- Check to see if the submittal is stamped as "reviewed" by the contractor, indicating he has examined it prior to sending it to the architect. Contractors are prone to stamp submittals as "received," and relay them immediately to the architect. The architect should insist that the contractor perform his own review of submittals.

- Compare the submitted product to the one(s) specified.

- If the contractor is proposing a substitution via the submittal, reject the submittal and refer the contractor to the substitution requirements section of the specifications (if appropriate). Substitution responses are covered elsewhere in this chapter.

- If the submittal shows the specified product, the architect must make sure it is correct in all regards (some products have the same numbers, but different sizes or colors).

- If the contractor asks for a verification of quantity, the architect should only do is if he is absolutely sure of the quantity and there are only a limited number of instances. Otherwise, just note: *Provide sufficient quantity in accordance with construction documents,* or *GC to verify.*

- If the product relates to any engineering discipline, the architect should check it for reasonable completeness and relay all copies to the proper consultant for his review.

- If the product submittal is incomplete, or information is not clear, mark *revise and resubmit* and return the specified number of copies to the contractor. The architect must be specific as to what additional information his is seeking.

- If the product submittal is clearly inappropriate, or is a substitution request without sufficient information, the architect should mark *rejected* and return to the contractor. The architect should be as specific as possible about why he is rejecting the submittal. In returning the rejected submittal, the architect must ensure that the proper chain of submission is followed so all parties are aware of this action.

- If the product submittal is correct and clear, the architect should mark it *approved* or *approved as noted*, and return it to the contractor.

- The architect should review his submittal stamp language with his professional liability insurance carrier to ensure that it is sufficiently clear and complies with the PL carrier's requirements.

- The architect should return the correct number of submittals to the contractor, retaining two sets for the office (one for himself and one for the owner). In the case of electronic submittals, he must ensure the copy list includes all parties in the submission chain. Architects should always copy the owner, so he is aware of which products have been approved on his project in sufficient time to raise an objection if he has a concern.

APPLICATIONS FOR PAYMENT

Applications for payment are how a contractor requests money from the owner. That part is simple enough, but reviewing and approving applications for payment can be a complex process, especially if the contractor is performing poorly or is well behind schedule. Even on a smoothly-running project, owners are particularly wary near the end of construction, wanting to ensure that they are holding sufficient sums of money to guarantee the project scope is completed, the punchlist is resolved, and credits due to them are paid. Money, in the eyes of most owners, is the lubricant to ensure that the contractor drives the project to completion.

As a design professional, the architect's role is to assess fairly the progress claimed on the application for payment against the work he observes in the field. In most lump-sum contracts, the contractor makes his case for payment by stating his percentage of completion against a schedule of values approved early in the project. This schedule, created before the start of construction, defines the basic stages of the work, usually by Construction Specifications Institute (CSI) division, with a dollar-value assigned to each portion. The contractor will estimate the percentage of work he has completed for each line on the schedule. That percentage, multiplied by the value, is the total amount he has earned to date. Once retainage is subtracted for each line, the application shows the total amount due for that line item.

All of these line items are totaled, and carried over to the front sheet of the application (see Figure 6-2 for a partial view of an application for payment form).

In an efficient manner, the application's front page should summarize the original value of the contract, the total value of change orders, the amount of retainage, the amount due for the current pay application, and the amount of value remaining to be paid. The architect should stipulate in the general conditions the form of the contract and application for payment. Forms published by the American Institute of Architects have become the standard for payment applications. AIA form G701® is the application front page; AIA form G702® is the schedule of values page. Even other non-proprietary forms follow the same basic format.

Application for Payment Strategies

A design professional's strategy for avoiding liability in managing pay applications begins with the schedule of values. Contractors create this schedule, but it must be approved by the architect or engineer. Contractors typically employ a simple strategy called "front-end loading" to get money flowing into their coffers quickly. Front-end loading is characterized by using heavier than normal mobilization, demolition, general conditions, site grubbing and clearing, and similar early project costs to establish working capital early in the project, often in excess of the actual costs incurred. In defense of the contractor, there are significant start-up expenses in mobilizing to start construction. Requirements such as a field office, permits, bonds, builder's risk insurance, and temporary power add up quickly. A general contractor is certainly entitled to reflect these costs on a schedule of values and seek payment for them. The architect should keep in mind as well that on a typical project a contractor will perform work for 30 days, make his first application for payment, and typically wait another 30 days for payment. By the time a general contractor sees his first payment--approximately 60 days into the project—he has made a financial commitment to the project and the owner is, in the jargon of payments, "deep into the contractor's pocket." While all of this argues for leniency in reviewing the schedule of values and allowing the general contractor some leeway in invoicing early for legitimate start-up expenses, the contractor is required to front these costs if he is unable to negotiate a start-up payment from the owner. The owner may also find that the values for early construction items appear heavier than he would expect. Clearing and grubbing of the site, for example, is legendary in construction circles as an extraordinarily expensive endeavor. This is a simple example of the contractor billing heavily early in the project to establish some cash flow and reduce his financial risk. Although some leeway is appropriate, if the architect's instincts tell him the values are excessive and the contractor has overreached, he should require the contractor to document why some of his line items are so high. If construction division bid responses *(See Chapter 4)* are required by the contract, the architect has this information available to use as a reference in challenging excessive front-end loading. Otherwise, he can refer to other local projects of similar type or standard cost references as a resource in questioning excessive start-up costs by the contractor.

1. Original Contract Sum:	$	-
2. Net Change from Change Orders:	$	-
3. Contract Sum after Change Orders:	$	-
4. Total Completed to Date:	$	-
From Schedule of Amounts		
5. Retainage: 10%	$	-
6. Total Amount Earned (minus Retainage):	$	-
7. Amount Previously Paid :	$	-
8. Current Amount Due:	$	-
9. Balance to Finish (including Retainage):	$	-

Summary of Change Orders		
CO #	Purpose	Amount
Change Order Totals:		$

The undersigned Contractor certifies that the work included in this Application for Payment has been completed in accordance with the Construction Documents, that amounts owed under previous applications have been paid, and that the amount requested on this Application is now due.

Contractor Signature: _____   Date: _____

FIGURE 6-2
Sample application for payment.

When reviewing a schedule of values, it is important for the architect to ensure that the line-item descriptions are neither too broad nor too detailed. This may sound contradictory, but the rationale is important: overly broad categories make it difficult to accurately assess the percentage of completion. A single line item for "concrete," for example, could consist of foundations, site paving, curbing, or interior slabs. All occur at different points in the project, and the architect would be hard-pressed to accurately estimate the total completion percentage of a single concrete line item at any point. When these elements are broken out as individual line items, however, it is much easier to assess quickly whether a percentage claimed by the contractor is valid or not.

Conversely, a contractor who creates an elaborately detailed schedule of values is creating a potential review headache for himself and the design professional. Contractors normally value simplicity in billing as much as design professionals, but occasionally the architect will encounter a general contractor who has decided, either through successful past experience or intricate logic, that a more detailed schedule of values will somehow enhance his billings. Using the same concrete example from above, if the contractor breaks out line items for concrete accessories, site curbing, building apron curbing, grade beams, flat slabs, column footings, and exterior foundation walls, he is increasing the time and difficulty required for him to prepare the application and for the architect to review it. This contractor will argue that a more detailed breakdown enables him to invoice fairly for partial work performed earlier (site curbing finished in advance of building curbing, for instance). With reasonable professionals, however, there is rarely an advantage to creating a much more detailed schedule of values (see one exception below). The same benefit can be derived through a reasonable assessment of the percentage complete on a single, all-encompassing curbing line item. Ironically, it is often easier to overlook parts of the work in a more detailed breakdown than in a simpler line-item breakdown. A contractor intent on utilizing a more detailed schedule of values will sometimes find that he has forgotten to include some portion of the work. He is then dependent on the good graces of the design professional to allow him to recapture it by billing ahead in another line item, or accept the architect's argument that it must, by default, have already been picked up unseen in some hidden line item. Such discussions get mess, which is why it is better to keep the schedule of values relatively simple and clean.

Where this may not hold true is in public work. This book deals only with private projects, but occasionally non-public projects utilize public funds in which a third party has the right to review an application for payment before releasing his agency's funds. Where a public agency is funding a component of the project (sitework, for example), it may be worthwhile to provide a level of detail for that part of the work necessary to satisfy the agency reviewer that the pay application is appropriate. Governmental agencies often frown on percentage of completion estimates, preferring more detailed line-item breakdowns that enable them to pay nothing until the entire scope is complete and accepted. When the architect encounters this situation, he should work with the owner to assess how the funder wants to review the application for payment, and attempt to meet his needs. This may mean creating a sitework schedule of values that is longer than that for the remainder of the project, but so be it. Above all, the design professional should do his best to keep the money flowing for work that is completed and in compliance with the documents.

A common dispute in pay applications is a disagreement with the contractor on the relative percentages he has completed for various line items in the schedule of values. Although a design professional can certainly take the time to estimate the quantity of work complete in the field versus the scope represented on the drawings, in reality it is more art than science. Contractors often offer to prepare a "pencil copy," or draft markup of the application for the design professional to review at a job meeting. The contractor's project manager and the architect/engineer typically walk the site together to roughly assess the percentage claimed as complete. Sometimes, the challenge to a clearly heavy percentage will be answered with "We're pouring that tomorrow," or "That'll be installed by the time you get the final application." These are judgment calls on the part of the design professional. It requires a balance between the realization that the owner is normally 30 days ahead of the general contractor in paying for installed work and holds retainage as well, and the fact that the architect is responsible for protecting the owner's fiduciary interests regarding approving pay applications. The architect may draw a hard and fast line or a fuzzy one in terms of allowing a few days' leniency in including work on the application. Either way, he must be consistent and communicate his rules clearly to the contractor. No one likes a baseball umpire with a variable strike zone, and no one appreciates a design professional who keeps changing the rules in the middle of the game.

In completing the application for payment, the architect should limit his liability by reminding the owner and contractor of the limitations surrounding his approval, such as:

- The approval is not the result of meticulously detailed or regular on-site observations by the architect. It is the result of visual and periodic observations of the work as it has progressed.

- The approval was not the result of an investigation into areas that are the sole responsibility of the contractor, such as means and methods related expenses or general conditions costs.

- The approval was based on data provided by the contractor, but not on any information provided *to* the contractor by subcontractors or material suppliers.

- The approval did not include an investigation into how (or whom) the contractor paid with funds previously released by the owner through the previous application for payment.

One final note regarding applications for payment: No contractor takes kindly to having his money held. It is critical for architects to review and process payment applications promptly to avoid liability for delaying the contractor, or imperiling his ability to pay his subcontractors and vendors.

7

Under Construction—
Problems and Disputes

DISPUTES

Disputes on the construction site can stem from any number of sources. Typically, however, they can be traced back to disagreements over quality of the work, change orders or other monetary issues, field conditions (hidden conditions or weather), and construction document issues (see Figure 7-1).

Architects are front-and-center in most construction disputes. The owner and contractor look to the architect to be an impartial interpreter of the intent of the construction documents, and the owner views the architect as his representative in observing the quality and completeness of the work. Sometimes the lines between these roles are blurred. The architect can trigger a dispute when he questions whether the contractor is performing the work in conformance with the construction documents, or when he challenges either the validity or value of a change order.

The best practices for resolving construction disputes include a combination of: dispute prevention, flexibility, early dispute intervention, use of alternative dispute resolution methods, and a predetermined plan as to how disputes will be handled. The importance of problem-solving, as opposed to adopting an adversarial attitude in the process, cannot be emphasized enough.

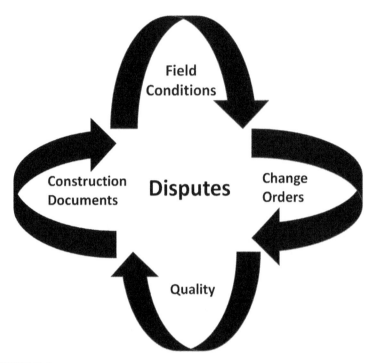

FIGURE 7-1
The origins of disputes.

The various types of dispute resolution methods used in construction disputes are:

- Litigation
- Arbitration
- Mediation
- Expert determination
- Conciliation
- Specialized large-project methods

Following is a summary of each of these methods of dispute resolution, beginning with the least desirable.

Litigation

Litigation usually refers to the process of submitting matters for resolution to the courts of the state, or in some instances, the federal court system. Lawsuits are endemic in the construction industry, and are filed for reasons as varied as the size, scope, and nature of the types of projects from which they stem. Some lawsuits are quickly resolved, with a minimum of rancor and cost (rarely), while others mutate into protracted, expensive, and risky litigation (typically).

Construction litigation can arise from any number of sources, including delay claims, construction defects, professional liability allegations (errors and omissions), mechanics' liens, personal injury, property damage, and insurance, subrogation, or performance bond claims. Construction disputes often involve arcane technical issues in specialized fields, such as critical path scheduling, geotechnical engineering, structural engineering, building code interpretation, material failures, defective work, performance issues, professional standards of care, or national reference standards.

The fact that a dispute reaches the point where one party is willing to commit the substantial time and money required to litigate a solution to it means that the problem has one or more of the following characteristics:

- The amount in dispute is large, relative to the size of the companies involved.
- One or both parties are unwilling to compromise, or discuss a conciliatory resolution to the problem.
- Construction has been halted, has already ended, or resolution of the issue will not affect the completion of the project.
- The problem is complex and multi-faceted. It involves a number of issues and parties, to the point where conciliation is not likely and where fault is likely to be shared among several parties.
- The issue may be resolved quickly if it is subject to legal motions that compel a resolution.

Litigation is the least desirable means of resolving a construction dispute. It is usually incapable of providing quick resolutions to disputes that occur during construction, and is invariably costly and time-consuming to pursue.

Arbitration

In this method of dispute resolution, the parties agree (usually in the original contract) to private dispute resolution. At the time one of the parties demands arbitration, they will also agree on an arbitrator, or a panel of arbitrators, knowledgeable in construction issues. Arbitration proceedings are touted as shorter than litigation, though they are still conducted in a formal setting where the arbitrator requests filings from each side and questions witnesses in the dispute to gain an understanding of the issues affecting it. Because of the complexity and value of arbitrated matters, lawyers are typically involved.

Depending on the provisions of the agreement between the parties, the arbitrator's decision may or may not be binding on both parties. If binding, there may be a limited right of appeal on questions of law, procedure, or the fairness of the process (such as bias, prejudice, or interest of the arbitrator). The arbitrator's award can be enforced in other jurisdictions.

Arbitration has fallen out of favor as a preferred dispute resolution technique because of the time involved in obtaining a decision, the cost, and the perceived tendency of arbitrators to evenly divide the cost of the resolution in particularly contentious disputes, deciding on roughly equal responsibility for each side in order to avoid the harsh results that may occur to either side if it loses on all counts.

Mediation

Mediation is a fundamentally different approach to resolving construction disputes. In litigation or arbitration, a judge or arbitrator will weigh the legal rights of each party in a dispute and decide who is responsible, to what degree, and what amount they are responsible for paying to resolve a dispute.

Mediation, on the other hand, is a voluntary negotiation between the parties in a dispute, aided by a neutral third party (the mediator). The mediator's role is to assist the parties to find their own solution. He cannot impose a decision, though he can make suggestions and cajole both parties to bring them to an agreement. The most recent version of the American Institute of Architects' standard forms added a mediation requirement prior to binding arbitration for all disputes.

Mediation has become popular as an alternative to arbitration because of its speed, lower cost, and ability to promote conciliation over an imposed decision.

Expert Determination

Some disputes are of such a nature that they can be resolved through objective testing or the opinion of an expert. Environmental and material testing issues are examples of problems in which both parties may voluntarily conclude that they will accept the judgment or report of an independent agency. Common examples of this type of situation occur with the quality of installations of various types of finishes. An owner may be dissatisfied with the appearance of a quarry tile installation, for example, claiming that the grout color is inconsistent or that the tile joint spacing is erratic. If the contractor and his sub refuse to accept the architect's opinion, the architect may turn to the reference standards for quarry tile installation he cited in his specifications. For a fee, the professional association used as a reference may send a representative to review the installation and offer an opinion as to whether the installation in question meets the reference standards. If both owner and contractor agree to accept the judgment of this individual, then the dispute can be solved through the use of this expert's determination.

Conciliation

Conciliation is the act of two parties in a dispute voluntarily agreeing to a resolution. Conciliation is, far and away, the most commonly used method of resolving construction disputes. It is also the most desirable, since it offers the benefits of speed and lowest cost.

Conciliation requires reasonableness and flexibility on the part of owner, architect, and contractor. It may require two or more parties in a dispute to accept some responsibility for causing a dispute, and hence some responsibility in contributing to its resolution.

The Engineers Joint Contract Documents Committee (EJCDC) standard forms require good-faith negotiations for 30 days before other means of alternative dispute resolution are undertaken.

Large-Project Techniques

Two techniques have arisen that have special applications in larger, complex projects involving sizeable organizations: step negotiation and dispute review boards.

Step Negotiation

Step negotiation is a formal technique of conciliatory dispute resolution that generally requires the individuals directly involved in the dispute to seek resolution through direct negotiation. If a resolution is not reached within a set length of time among the principals involved in the construction, the dispute rises to the organization's next level as defined in the process. This step process continues until it reaches the senior levels of each organization. The Design-Build Institute of America (DBIA) standard forms require step negotiation, followed by mediation and binding arbitration, as the preferred method for resolving construction disputes.

Step negotiation is a useful tool for large projects involving large organizations. A primary benefit is that any problem that reaches senior management is less likely to be encumbered by emotions that may preclude settlement of an issue among the lower parties more familiar with the problem. The detriment, of course, is that those remote from the problem may be less familiar with the intricacies involved in resolving it.

Yet another advantage of step negotiation may be the organizational reluctance of lower level managers to pass the buck to their senior managers, who may be unhappy to spend the time required to resolve these issues. This reluctance may persuade the lower managers to be more amenable to reaching an agreement with each other.

Dispute Review Boards

Dispute review boards typically consist of three neutral experts who visit the site to periodically monitor progress and stay aware of potential problems. When requested by the parties to an agreement, the board conducts an informal hearing of a dispute and issues an advisory opinion that the parties can use as a guide for further negotiations. Dispute review boards are a more formal method of guided negotiations. They are beneficial as a tool to help each party gain an objective perspective at the onset of a problem. They are less useful in that they have no direct part in the negotiations and possess no real authority to push both parties toward consensus.

POOR WORKMANSHIP

One myth should be disposed of immediately. Successful contractors do care about the quality of the work they perform, and usually work diligently to correct work of poor quality. They have learned that a reputation for poor craftsmanship will dog a contractor long after the project is complete, and long after payment disputes or change order issues are forgotten. Contractors, at least the ones who have been in business for a long period, are genuine professionals and care about their reputation as much as any other professional. In that sense, the title of this section is misleading. Contractors refer to their work issues of workmanship in a more positive light; they call it construction quality control. Larger construction firms have quality control personnel on their staffs. Their job is to implement programs and perform job site checks to ensure that their in-house personnel and subcontractors are performing work that exceeds the minimum quality standards demanded by the specifications and working drawings.

Most contractors do not have the benefit of separate quality control staff, and rely on their own sense, borne from whatever experience they possess, to know what represents quality construction and what represents something less. Contractors do employ quality control measures, either formally or informally. A formal quality control plan for a contractor addresses the entire spectrum of issues affecting the construction of the project, including:

1. *Organizational responsibilities and duties of personnel*
   a. Project manager
   b. Field superintendent
   c. Quality control manager
   d. Safety manager
2. *Submittal procedures*
   a. Approval procedures for shop drawings and submittals
3. *Three-phase inspection process*
   a. Preparatory inspection
   b. Initial inspection
   c. Follow-up inspection
4. *Testing and certifications (see Chapter 5 for more information)*
5. *Tracking and record-keeping procedures*
   a. Requests for information
   b. Submittals
   c. Change order requests
   d. Applications for payment
   e. Miscellaneous correspondence
6. *Coordination meetings*
   a. Contractor and subcontractors
   b. Owner and architect
7. *Schedule conformance*
8. *Inspections*
   a. Punch-out
   b. Pre-final
   c. Final acceptance

In dealing with quality control concerns in the field, architects face three different types of situations, including:

1. ***Concealed or covered-over work:*** This situation occurs when the pace of work does not allow the architect to view an installation before it is concealed. In most cases this is not due to a contractor's desire to conceal deficient or non-compliant work, merely his desire to push the schedule. The fact that the architect's site observation visits occur at infrequent intervals also causes him to miss parts of the progress of the work. Where the architect is concerned with an aspect of construction (wall framing, for instance) that he wishes to observe before it is covered up, he should alert the contractor to this fact and make himself available to visit the job site when the work is completed to avoid disrupting the contractor's (and owner's) schedule. For less important issues, he can ask the contractor to take photographs of the work before it is covered to document that it was performed properly. Another option for the architect is to request an as-built drawing from the contractor to document an installation that cannot be viewed and is impractical or expensive to uncover for inspection.

2. ***Procedure disputes:*** Architects will sometimes observe installation procedures that they do not approve of and that do not conform to the reference or quality control standards published in the specifications for the project. While architects should normally grant contractors some leeway in achieving satisfactory results through alternative means, where a contractor or his subs are clearly not following well-known criteria, the architect must voice an immediate concern in the field, and follow-up in writing with the contractor. In some instances, the architect may be observing the improper installation in advance of the field superintendent, who will correct it once he becomes aware of it. In other cases, the field superintendent will defend his subcontractor's method of installation as appropriate, leaving the architect to state in writing why the method is not in conformance with published standards or common practice, and state his concern for the ultimate quality of the work. The architect should have no hesitancy in reminding the contractor of the risks associated with

not following the proper procedures, and that the resulting work will not be in conformance with the requirements of the specifications.

3. ***Latent defects:*** Latent defects are those that become evident well after the work is complete. The lack of levelness of a ceramic tile floor may not become evident, for example, until the grout is in place and the floor is cleared. Similarly, poor drywall finishing on an expanse of wall may not be evident until the light is raking across the wall during the morning or afternoon; or a boiler pipe may leak after several weeks of use. These situations can be difficult to correct after the architect has previously accepted the work, the contractor has removed lifts or other equipment necessary to correct the work, and paid the subcontractor responsible for performing it. These instances are exactly why retainage is required. Retainage provides the owner with the financial leverage to compel the contractor to correct work that subsequently reveals itself to be deficient. A contractor truly interested in providing a quality product to his client will not argue over clearly deficient work. Retainage, withholding of final payment, and the warranty guarantee are tools the owner and architect can employ to address those times when the contractor is not willing to voluntarily correct a latent defect.

HIDDEN CONDITIONS

Hidden conditions in construction occur in areas where remediation or repair work is suspected but cannot be seen, such as behind walls, in interstitial areas, and under roofs or behind facades. If the problem is extensive, the architect may decide that total removal and replacement is the most cost-effective means of dealing with it. The problem with explaining hidden conditions is that often they are not truly hidden. Water stains, odors, finish deterioration, or other indicators can demonstrate the existence of a problem. What is hidden is the extent of the problem. Architects will sometimes try to shift the problem of identifying and correcting hidden conditions to the contractor through the use of all-encompassing notes such as:

- Contractor responsible for correcting all water damage discovered during construction
- Contractor shall demolish and reframe all wall construction found to be deficient.

Use estimated quantities and unit prices to cover some of this scope in the bid documents, but in doing so the architect must make assumptions about what work will be required. Unfortunately, much of it may be little more than educated speculation. The important attribute, however, is that he is building value into the construction documents to offset whatever the actual work turns out to be. This is useful for two reasons: 1) It is good client management for an architect to be able to show he attempted to manage a problem of unknown dimensions, and 2) If the architect has even a portion of the solution covered, the owner receives the benefit of the bid environment in determining the pricing (see the Risk Hazard Flags).

Hidden conditions are particularly difficult to deal with when they involve environmental hazards. On occasion, these hazards are in plain sight, as when asbestos binder is discovered in plaster or lead-based paint is discovered on woodwork. Although these hazards were indeed *hidden* in the sense that they were unknown to any party prior to their discovery, they could have been uncovered by vigilance or testing. Because they were not, the owner will likely suffer a delay in his project, as well as extra mobilization or general conditions costs that could have been avoided if the problem had been discovered before the contractor had commenced work.

 **Risk Hazard Flags**

Common Hidden Condition Mistakes

- Ignore signs indicating a problem
- Use general notes to make the contractor responsible
- Assume worst case and replace everything
- Assume all hidden work will be change order work
- Tell owner he is responsible for hidden conditions

Architects are not usually held responsible for the recognition of environmental problems on a project (and are usually contractually excused from such investigations), but in an existing building architects should recommend to the owner that basic environmental testing be performed as soon as renovation is contemplated. The budget and schedule consequences of remediating environmental hazards can be so severe that not testing early for these potential problems is an unwarranted risk.

CHANGES AND CHANGE ORDER REQUESTS

Change order requests (also referred to as pending change orders) are notices from the contractor that he is claiming an increase in the contract sum for work that he believes is not part of his contract with the owner. Change orders are approved change order requests, where the architect and owner have agreed that the contractor is entitled to an increase in the contract sum (and sometimes additional construction time as well). An approved change order may consist of any number of change order requests for a variety of purposes. A contractor tracks change order requests on a separate log showing the date they were issued, the purpose, amount, and resolution (denied, disputed, or approved). Once they are approved, change orders are added to the contractor's periodic applications for payment and schedule of values. The contract sum is increased (or decreased) by the amount of the change order, and they are tracked by percentage of completion, as with any other item on the schedule of values.

Change order requests can occur in the form of extras or credits. Extras are demands for additional compensation; credits are acknowledgements that the owner should be rebated money for work in the contract that was not performed (see the Risk Reduction Tools for a list of change order response tools).

A change order request is exactly that—a request. For change order requests not initiated by the owner, the contractor must meet a burden in three areas before the architect can reasonably say the request should be approved by the owner:

Risk Reduction Tools

Change Order Response Tools

- Is support information complete?
- Is the work indicated in the CDs?
- Is the work implied in the CDs?
- Was related work shown in the CDs that warrants a credit?
- Is the work a customary part of the installation?
- Is pricing reasonable and customary?
- Is a schedule extension warranted?
- Are additional general conditions included in the change order value?

1. He must prove the work is not reasonably part of the contract for construction.

2. He must prove the necessity for the additional work to be performed.

3. He must document the value of the additional work, or how that value will be determined in the future.

Each of these requirements must be fairly assessed by the architect on behalf of the owner. Of all the issues in construction that pose the potential for owner/architect conflict, nothing can hold a candle to change orders. Except for change orders resulting from scope changes and differing site conditions, a change order request represents a statement by the contractor that the construction documents were lacking in some way. If the change order represents a cost for ordered products that had to be returned or for installed work that had to be removed, the architect may face accusations from the owner that the architect's negligence cost him money that he otherwise would not have had to spend, money that is essentially wasted. Even if the change order represents the cost of a product that was inadvertently left out of the documents, but that the owner expected to have been a part of his project, he may demand that he receive it at no additional cost.

Compounding this problem is the phenomenon accepted in construction circles of the credit/extra paradox: credits are devalued, extras are overvalued. So even for a situation where the architect and the owner agree that a desired element was not shown in the construction documents and not reasonably a part of the contract for construction, the owner may still claim that the architect cost him money by denying him the right to have this product priced in the more favorable *bid climate* as opposed to the clearly unfavorable world of change order pricing.

Clearly, reviewing and responding to change order requests represents the high-wire act of construction administration. An architect who has strong documents to defend against change orders, and does so effectively and consistently, will earn the grudging respect of contractors and the continued patronage of his clients. Those who do not will still be admired by contractors, though for different reasons. Listed below are the basic concepts to remember in reviewing change orders, centered on the three areas of proof required of the contractor.

Is It in the Contract?

A change order request is most often based on the argument that the work in question is not part of the contract for construction. The contractor will argue that the work is not explicitly called for in the plans and specifications, and therefore he should be paid extra for performing it. He may also argue that the work was shown in a contradictory manner on the documents, and he assumed the less expensive route as part of his bid. To prove this, the contractor is essentially stating a negative: He cannot find the required work in the documents. The contractor's argument is almost always based on explicit evidence: No detail, note, or schedule indicates this work. The architect can defend himself by finding indications to the proposed work, either explicitly or implicitly (see the Risk Reductioin Tools).

Risk Reduction Tools

Document Interpretation Tips

- Cite all related information in the drawings and specifications, even where contradictory
- Reference the specifications
- Reference quality standards if applicable
- Be pragmatic in interpreting implied work
- Ask the contractor what he assumed during bidding

For change order requests that allege no indication of the work claimed in the change order request, the architect has the following options to defend himself:

1. ***The work is implied:*** Though not explicitly shown, the work is implied in the documents and should have been considered included via a reasonable review of the documents. Example: Circuiting for an electric heater is left off the electrical plan but is clearly implied by the fact that the heater is listed in the appliance schedule.

2. ***The work is an integral part of a finished project:*** Even though not indicated, the work in question is such a fundamental part of finished construction that a reasonable contractor would not have excluded it. Example: Metal coping is not clearly indicated on a parapet wall, but no reasonable person would argue that a masonry wall could be left exposed to the elements without protection.

3. ***The work is included is in the specifications for the project:*** Example: Even though an exterior door hood was not shown on the elevations, it was called out clearly in the specifications. While the quantity of the hoods may be in question, the contractor was aware that the product was in use on the project. If the architect's front-end specifications

contained a clause stating that "all ambiguities or inconsistencies in the documents will be decided in favor of the *higher quantity or better quality* product," then even this debate can be settled quickly.

Construction documents are difficult to coordinate thoroughly, so it is not difficult for a contractor to find contradictions in them. When a contractor bases a change order request on the fact that the documents show contradictory features, the architect defends himself in the following three ways:

1. ***Preponderance of evidence:*** Even though there is contradictory information in the documents, most of it points to the more expensive solution. Since specifications represent the most specific portion of the documents, it is quite helpful for the architect to be able to point to a specification item to bolster his case.

2. ***Implied work:*** Even if the specification called for the lesser work, the space allowed, circuiting, and all other indicators in the documents pointed to the more expensive work.

3. ***Clarification:*** Documents are not perfect. Where a clear contradiction occurred, the contractor owed an obligation to the owner to seek clarification through the architect. He should not have made an assumption; he should have sought the answer via RFIs or questions during bidding.

Is the Work Necessary?

After proving the work is not part of the contract for construction, the second test a contractor must pass in claiming a change order is to prove that the work is indeed necessary. This is usually an easy task for the contractor, but it is also an area where the architect's skills and creativity can bear fruit in saving the owner money. Example: The electrical schedule shows three-phase equipment on what is otherwise a two-phase project. The contractor requests a change order for costs associated with installation of a phase converter to power the three-phase equipment. Since most of the equipment in question has not yet been ordered, the architect works with his engineer to specify new two-phase equipment. The change order is reduced to a small restocking charge for the three-phase equipment that is already on site.

In other instances, the architect may find that some skillful redesign may eliminate, or lessen the cost of, the change order. Contractors and subs can also be helpful in suggesting alternatives that will cost less than the submitted change order. They usually wait to be asked, however, so the architect should not be shy about expressing concern over the cost of the change order request and seeking suggestions to lessen it. The architect should lead the value-engineering effort among his consultant team and the contractor if it helps to mitigate the costs of the change order.

Document the Value of the Work

If the contractor passes the first two tests for a successful change order request, he must then document the value of the increased work. For most work, this consists of a detailed subcontractor's proposal with material and labor values, broken down by hours and the rates charged per hour. If a significant material cost is included, the subcontractor should include the quote from a supplier documenting this cost. Architects can check the hourly rates and amount of time claimed for the work against standard cost estimating references, keeping in mind that specific job conditions can vary the labor time considerably.

Where change order work is so critical that it must begin immediately (or is already in progress), contractors may offer to work on a time-and-materials basis. Architects may suggest a lump sum or not-to-exceed (or *upset*) limit to protect the owner in these instances. As verification, the contractor must track time of the subcontractors or workers via daily job tickets signed by the contractor's field superintendent and make available invoices to document all materials. Labor rates should be agreed upon ahead of time, with additional stipulations regarding overtime amounts and rates. Unit prices and allowances should be referenced in change order requests whenever they are relevant.

C.A. Anecdote

Tale of the Frozen Condensate

The Problem

Joe Morris took his time getting into the office on a frigid December morning. He had no meetings scheduled, and looked forward to a quiet day reviewing closeout documents on a 95-unit condominium project his firm had recently completed. He had barely sat down in his office when one of his young associates rushed in.

"Big problem at Brookmeade Condos," she said. "The furnaces aren't working."

"This isn't our problem. Call the contractor and ask him to get his mechanical sub out there to check it out," said Joe.

His associate explained that the contractor and his sub had called her from the site. Although only fifteen units were currently occupied, the furnaces in every unit had shut off, and the residents were cold and angry. A quick conversation with the mechanical subcontractor had revealed the problem to her.

Their mechanical engineer had specified high-efficiency furnaces that produced only a slight flow of condensate in the winter. The condensate line exited a few inches above grade, allowing the slow drip of condensate in freezing weather to build up a perfect ice cone beneath each line, eventually blocking it and causing the furnace to shut off.

Joe leaned over his desk and put his head in his hands. "They need condensate pumps," he said. At $250 per unit for the fix, he had suddenly developed a $24,000 headache.

The Resolution

Joe met his client, the contractor, and the mechanical sub at the site that afternoon. He broke the tension immediately when he told the group: "That wasn't one of our cleverest details." Most insurance companies advise against making any statement of liability, but Joe felt it was critical to avoid a protracted argument over responsibility and gain the contractor's assistance in solving a critical problem as soon as possible. The subcontractor, happy not to be the subject of blame, agreed to a fair rate for the labor to install the pumps, as long as the owner purchased them. Joe's client was initially unhappy about paying for the installation of 95 new condensate pumps. Joe explained that the gravity design was an attempt to provide him with the lowest-cost solution, but it clearly was not working out. The owner was upset not only that he needed to pay the change order, but that he could have saved money if the architect had anticipated the need for the condensate pumps and included them in the bid documents. After some difficult discussions, the owner finally conceded that the pumps would otherwise have been required, and that he was not entitled to the "unjust enrichment" of asking the architect or his liability insurer to purchase them for him. He did insist that the architect compensate him for an amount, which they negotiated, to represent the excess cost of installing the condensate pumps under a change order.

In all approved change order requests, unless specifically waived, the contractor is entitled to reasonable overhead and profit, usually specified in the agreement with the owner.

Nuances of Change Order Requests

Change orders are generated for a variety of reasons. Like ice cream, they come in different flavors, some of which are more palatable than others.

The Half-Hearted Change Order Request

Sometimes a contractor does not agree with a change order request he has submitted to the owner on behalf of his subcontractors. An architect can almost sense it in the passionless defense the contractor mounts to challenges, and the time in which he responds to requests for additional information. These are probably instances in which a contractor is passing on to the owner a change order request from one of his subs or suppliers that he does not necessarily feel is valid, but agreed to submit nonetheless. Like so many other business relationships, this is a path of least resistance in assuaging a disgruntled sub by at least asking for additional compensation on his behalf. The conversation between contractor and sub about the matter usually runs something like: "I'll ask, but if it's denied you'll have to eat it." Contractors will sometimes offer to waive their overhead and profit on this type of change order as a way to help sell it to the owner. Aside from the signals stated earlier, there is no way for the architect to recognize requests of this type on the front end. The best test is that if the architect feels a particular change order is subcontractor-generated and appears to be questionable, an experienced contractor would probably agree with him. Despite its origins, such a request must nevertheless be dealt with seriously and decisively, with a proper record of the resolution kept in the event the contractor renews the claim later.

The Owner-Initiated Change Order

Architects should always track owner-initiated changes. It is easy for owners to look at the change order tally on the pay application at the end of the project and complain to the architect that woefully deficient documents were the cause of these extra expenses. Owners tend to forget about all the change orders they initiate during the course of a project. It is worthwhile for the architect to keep a running tally throughout the job of the three sources from which change orders originate: owner-initiated (scope change), document-based, and hidden or field-condition based. The architect should also create a record of those change order requests that are based on alleged errors or omissions in the construction documents, particularly noting those where the owner has agreed to pay them despite protests from the architect that the change order is not warranted. The architect should object to payment of each change order with which he disagrees.

Request versus Reality

Any change order that is based on construction document deficiencies is a hit to the architect's image. This cannot be helped. Still, on any project where the change order requests occur, the architect is actively involved in defending the owner against unwarranted or excess claims for money, and working with the contractor to find less expensive ways to meet the owner's needs. Architects should track this progress, if for no other reason than defensive purposes. The architect should keep a spreadsheet tracking the amount of the original change order request, why it occurred (owner, architect, unforeseen), and the final approved amount. Odds are, the architect will be able to show that he saved the owner some amount of change order money through his diligence and hard work. It is a useful piece of information to share with an owner complaining about the cost of change orders at the end of a project.

The Change Order Offset

Change orders, of course, come in two forms: extras and credits. Contractors often forget the second type in their rush to claim a change order for any work that is not explicitly indicated in the construction documents. In reviewing a particular change order request, the architect should always look for any work associated with the change order that was clearly included in the contract and does not now have to be performed. The value of this work is owed to the owner as a credit, and will offset the extra work. Example: *The electrical subcontractor claims a change order for installing new deck-mounted light fixtures in lieu of the originally specified pendant-mounted fixtures. He issues a credit for the value of the original fixtures, but not for the extra labor saved in mounting the new fixtures directly to the deck.* The owner, interestingly, is not required to consider the credit offset in seeking compensation from the architect's professional liability carrier for errors and omissions claims. He may claim the gross value of the change order.

Defeating the Dubious Change Order Request

Some contractors follow the axiom with change order requests that "there's no harm in asking." They will submit requests for issues large and small, figuring that some percentage will be paid and will represent monies they otherwise would never see. As previously noted, contractors will also submit subcontractor or supplier requests that should never see the light of day. These requests can often be identified by the lack of documentation that accompanies them. There may, for example, be little in the way of explanation why the change order request is justified, with perhaps as the only backup a subcontractor's proposal with a lump sum number and no breakout for labor or materials. The architect should not immediately shoot down this fat target. Deficient though it may be, immediately rejecting it on behalf of the owner may simply move it into the contractor's disputed column, where it will reappear at the end of the project as an item for negotiation and settlement.

A better approach may be to return it to the contractor with a memo explaining why it is not in condition for review, and what additional information is necessary to make it so. Given that this change order request represents a quick and half-hearted attempt by the contractor, it is also an opportune time for the architect to fire a healthy warning shot that this request will face a tough fight. The best way to accomplish this is to hint at a multi-front attack that will require considerable time and effort by the contractor to overcome. The implied message: *Is it worth it?* A memo with this purpose could include the following elements:

- The work in question is included in the documents.
- Even if it was not, it would be implied as a customary part of a completely constructed work.
- Did the original proposal to the contractor specifically exclude this work?
- What did the contractor assume in his bid as an alternative to this work? Where is the credit for this assumption?
- The lump-sum proposal is unacceptable as backup to the change order.
- Acceptable backup includes references to the plans and specifications, and contains material and labor breakdowns, including hours and rates.
- Quotes from suppliers are necessary to document the material cost.

OWNER-FURNISHED ITEMS AND OWNER'S OWN FORCES

Owners often have specialized equipment and installers they wish to use in construction of their facilities. This occurs most often in retail or tenant-improvement work, where national chains have packages of finishes and furnishings they use in prototype facilities, and may even have selected installers they prefer to work with. The most desirable method of handling these situations is a defined ending point for the contractor, at which the tenant or owner accepts the space for his uses. The contractor vacates the work and the *owner's own forces* enter and carry it to completion.

Rarely does it happen this cleanly. More often, the owner or tenant supplies various types of equipment and furnishing packages for the contractor to install under his permit. This is useful in the case of electrical, mechanical, and plumbing trades, where it would be expensive and tedious for the owner to arrange for separate subcontractors to handle installing his equipment. Where this "blending" of owner and contractor work occurs, there is always the potential for problems. Here is a menu of issues affecting owner-furnished items and owner's own forces:

- Delivery or unloading responsibilities
- Storage location and conditions
- Installation responsibility
- Plumbing and electrical system compatibility
- Permitting issues
- Insurance protection
- Security for stored or installed products
- Union versus non-union labor
- Contractor coordination of owner's forces
- Schedule delays or complications
- Damage to or removal of new construction by owner's forces
- Damage to owner equipment or furnishings by contractor's forces
- Testing and warranty of owner-supplied equipment
- Delivery of damaged goods, or on-site damages.

To eliminate as much potential for misunderstanding and disputes as possible, the architect should work with his client to define clearly and specifically the requirements of the contractor in this area. In cases in which the owner is supplying equipment for installation by the contractor, for instance, the architect should try to provide a detailed list of the equipment, including power and plumbing requirements. The specifications should state who is responsible for accepting and unloading owner-furnished items, how they are insured, and how they should be stored. Labor and permitting issues regarding owner's own forces need to be discussed between owner and contractor prior to completion of the owner/contractor agreement. In particular, any issues revolving around union and open-shop labor should be resolved early in the process.

Architects particularly rely on specification language to define the responsibilities of the owner and contractor. Some examples of specification language affecting owner-furnished items or owner's own forces include:

The owner will provide certain items of equipment and furnishings as indicated on the drawings. The contractor shall be responsible for accepting delivery, unloading, and installing these items, including but not limited to: physical installation, utility rough-ins and final connections, and testing of installed units.

The owner will remove or relocate existing furniture and equipment from areas in which the contractor is working. The contractor is responsible for notifying the owner's representative not less than two working days prior to starting of work in areas from which furniture and equipment is required to be removed.

Concurrently with the work of this contract, other contractors, suppliers, or owner's own forces will be working in close proximity to the contractor's forces. Contractor will be responsible for coordinating his work with that of other personnel under the base contract, shall make no claim for the costs relating to this work, or schedule delays relating to a failure to do so.

INTERPRETING THE CONTRACT DOCUMENTS

The architect has an interesting role during construction. Although he is contractually bound to the owner, he is also the independent creator of the construction documents. When, in the course of construction, the meaning of the documents is unclear or disputed by one side or the other, the architect is called upon to be an impartial *interpreter* of the documents he has created. In the normal course of practice most architects, being human, tend to side with the owner's belief that something the contractor claims is not in the contract is either implied or implicit in the documents. There are times, however, when owners will make unreasonable or unjustified claims for additional work by the contractor and look to the architect to lend support through his interpretation of the documents. In these instances, the architect must maintain his stance as a neutral party and issue an interpretation that honestly reflects the design intent in preparing the documents. One side in the dispute, and possibly both, will be unhappy. The architect, however, will have maintained a reputation as a fair and independent interpreter of his work and avoided allegations that he is biased, partial, or unfair.

8

Ending Construction

CONTRACT CLOSEOUT

Contract closeout is that wonderful time in the life of a project when it is all over (or nearly all over), when the contractor and owner are ready to close out the contract and declare their project at an end. It is a time of some peril for the architect, since the owner is relying on the architect to advise him whether everything owed to him under the contract has been provided. The end goal of all construction projects is the completion of the construction so that the owner is provided with the project he hoped for, the architect designed, and the contractor promised to build. The project must not only look like what was anticipated but must also perform as reasonably anticipated for its intended purpose This is often expressed in the concepts of substantial completion and final completion. These phases are differentiated in construction to recognize the degrees of availability for use and the fact that certain phases or parts of the work may be complete and are available for use while others may not. The designations have important implications for the project, and the date of the determination of substantial or final completion may impact final payments to the contractor, retainage, availabilities of contract warranties, entitlement to bonuses, or liquidated damages.

For the most part, the building construction is the least of the architect's concerns at the extreme end of the project. The problem instead is one of ensuring that a whole host of documents (see the closeout documents section later in this chapter) are submitted by the contractor, and that he performs a number of closeout actions necessary to transfer full control and responsibility for the facility to the owner. Closeout documents are difficult to obtain once final payment has been made. A burden is often placed on the architect to ensure that the closeout documents are complete and acceptable. As a general rule, the owner should expect all punchlist items to be completed before final payment and contract closeout. Following is a list of common closeout steps necessary at the end of a project. Many of these actions are routinely taken care of by the contractor (for he is contractually obligated to do so), but the architect does well to at least take the effort to confirm that they have been done.

- Determine from the contractor that he is complete and ready for a substantial completion inspection
- Perform the final punchlist inspection
- Confirm removal of temporary site facilities and trailers
- Review the final application for payment
- Review all final lien releases
- Confirm receipt of the certificate of occupancy
- Remind owner to confirm that property insurance is in place
- Confirm final cleaning was performed
- Confirm all site rubbish/debris was removed
- Review the as-built drawing submissions from contractor and relay them to the owner
- Review the operation and maintenance manuals and relay them to the owner
- Confirm all product warranties were received and relay them to the owner
- Confirm systems start-up and training was performed
- Confirm spare parts, touch-up paint, and additional material was left to the owner
- Confirm final keying was performed, and that key/security information is in possession of the owner

- Confirm all utility transfers, from contractor to owner, have occurred
- Assist in reconciling all outstanding change orders and claims, and in preparation of final change orders and releases.

When the project or a portion of it as contemplated by the parties is sufficiently complete to be useable by the owner for its intended purpose, it is said to be substantially complete. While the contractor will notify the architect of his belief that the project has reached this stage, contract documents usually require the architect as an impartial arbiter to be the decider of the actual status of the construction. One reason that the architect is put in charge of this decision is his familiarity with the design documents and the design intent, even though he may not have been the actual designer. Another reason is that although the architect is for so many purposes the owner's representative, in determining completion he is to be an impartial arbiter, since both the owner and the contractor, due to the financial implications of completion designation, may have a bias on whether the project has reached this status.

SUBSTANTIAL COMPLETION

Once the contractor suggests completion, the architect makes an inspection for the purpose of determining the date of substantial completion. He will typically be accompanied by representatives of the contractor and the owner so notes can be made as to minor deficiencies in the work which can be quickly corrected. These minor deficiencies are compiled into a punch list which is completed by the contractor and inspected again by the architect for overall completeness of the list. The fact that the architect may participate in the creation of the list of deficiencies should not be used to suggest that the architect is in any way responsible for the correction of the deficient items. Importantly, if the deficiencies are not of a minor and relatively infrequent nature, then the architect must ask himself whether or not the project is truly substantially complete or whether the contractor has declared himself complete to qualify for a contract bonus or avoid a claim of liquidated damages. To guard against the contractor inappropriately calling out substantial completion, the architect should include in the construction documents a provision for backcharge in favor of the owner where the architect must perform multiple visits when the project was obviously unready for the inspection.

The determination of completion is sufficiently important as to require attention to the details and memorialization. The notice to the architect, the punchlist for correction, the notice of completion of the punchlist and, finally, certification of substantial completion should all be in writing and no payments should be approved until this paper work is completed by the respective parties.

FINAL COMPLETION

Once the architect is satisfied that the contractor has completed the punchlist, and satisfied all other requirements under the contract, he is then in a position to certify final payment. This may be the last substantive chance the architect has to assure contract document compliance before approval of the final payment, less long term retainage, and the loss of the contractor's zeal to complete the project. In addition to the punchlist, the architect is obliged to ensure that the contractor has fulfilled other contract requirements (as stated earlier in this chapter), such as provided as-built drawings, turned over all operations and maintenance manuals, perfected warranties on appliances, equipment and building systems, and provided specified training for the owner's agents and representatives for electrical, mechanical, and plumbing equipment and systems.

Final completion can often be a difficult stage of the project to control as the contractors have already commenced or even completed demobilization and the owner may be busy using the building for its intended purpose. The key parties are not always focused on the tasks necessary to closeout. Nevertheless, it must be the architect who keeps his eye on proper documentation of the closeout which may provide the best protection against claims by contractors or owners. For this reason it is especially important to resolve as completely as possible open items of bonuses, liquidated damages, as-built conditions, compensation for additional work, allegations of delay, and any other unsettled issue at the completion of the project. Where possible, a final and all inclusive change order should be prepared by the architect and signed by relevant parties to resolve these issues and eliminate as much as possible the prospect of contract claims.

PUNCHLIST

The punchlist inspection is the one time in the project when the architect is called upon to take a detailed look at the quality and completeness of the work. Indeed, it is the only time (other than final completion) when the architect is called upon to make an inspection of the quality of the work. As discussed in Chapter 5, the architect's field observation activities are periodic and not intended to be detailed *inspections* of the work. They are general observations of progress at intervals and as such, represent little more than snapshots of the work at it moves toward completion (see the Risk Reduction Tools).

The punchlist inspection itself cannot be all encompassing. The scope of work on any project is too broad to be reviewed in one inspection trip, no matter the number of hours. By the end of the project much of the utility and services work is covered over, and structural details may be concealed behind drywall or other finished work. The timing of the inspection can be very important, both for

 *Risk Reduction Tools*

Effective Punchlist Techniques

- Organize the punchlist by building area and trade, keyed to the plan, and consistent with the CSI format
- State the problem location, type, and correction clearly and suc-cinctly ("Replace damaged supply air diffuser over work area")
- Use action verbs for work that is required (repair, replace, touch-up, refinish)
- Note items that are owner agent's responsibility
- Do not field-mark items unless owner and contractor agree

contractual compliance and to ensure that the appropriate work-force is available to correct any deficiencies that are found. For the punchlist, therefore, the architect is concerned with fairly assessing the quality and completeness in accordance with his construction documents, based on what he can observe, and creating a list of what needs to be done to complete the construction. He, or his consultants, may run water faucets, check the polarity of electrical receptacles, or tap on ceramic tiles to ensure they are bedded properly. He may do all of these things or none of them, depending on the project, but he cannot be responsible for verifying in this one trip that the contractor has met all the requirements of the construction documents. That remains the responsibility of the contractor. The punchlist inspection, whether thorough or cursory, offers no assurances to the owner that the contractor has done everything required of him. The contractor himself provides that assurance to the owner through the warranty. The architect should make sure the owner understands the purposes and limitations of punchlist inspections.

So what does the punchlist inspection accomplish? It is an outside review of visible and easily discovered items that affect the owner's immediate use of the facility, as well as a determination of what is needed to finish. Some typical punchlist items include:

- Drywall finishing
- Paint coverage and colors
- Finish product installation, fit, and trim
- Exterior closure elements: completeness and quality
- Roof coverage and general acceptability
- Operation of plumbing fixtures
- Light switching; number and type of light fixtures
- Operation of HVAC units and equipment
- Verification of sprinkler heads, exit signs, and emergency lighting
- Fit, finish, and operation of doors, windows, or glazing
- Casework or custom elements fit, finish, and operation.

In preparing the punchlist, the architect should follow an organized path through the project, listing items by space number, and then by trade within the space. When possible, he should be accompanied by the owner or an owner's representative, so they can view the work together and agree on the punchlist items. This will make it easier for both the field superintendent and the architect to check progress. He should be specific in his comments and use active verbs to communicate the corrective work that must be performed. A typical punchlist entry might look like this:

Space 201: Kitchen—Wall area beneath left side of window requires respackling and touch-up.

Some architects prefer to use small colored sticky dots to indicate areas that are listed on the punchlist. These can be helpful, but should be discussed with the field superintendent first. They generally are not necessary if the punchlist is specific enough, and some superintendents may be annoyed by seeing colored dots scattered around their newly-finished building. An architect should never use permanent markers of any kind to mark work he considers defective. This rule applies for two reasons. One, the architect may not be the final arbiter of defective work. If the owner accepts it, then the architect must accept it as well (unless it is a life-safety item). Permanently defacing a part of the work denies the owner this benefit. Also, contractors cannot always complete all punchlist items before owners occupy the space. Permanent markings, even on clearly defective work, devalue the quality and usefulness of the space for the owner.

Once the punchlist is delivered to the contractor (and copied to the owner), the contractor typically makes copies and distributes them to the affected subcontractors so they can begin work on their items. Contractors also review the list for punchlist items they wish to dispute on some basis. Typically, contractors may dispute a punchlist item for several reasons:

1. The work was performed or damaged by an owner's agent or others not under his employ.

2. The work was previously accepted by the owner or architect.

3. Problems with the work occurred after completion, and were caused by another party (owner's own forces, prime contractor, or non-participant in the construction).

4. The contractor believes the work is of satisfactory quality and in accordance with the contract documents.

The first three of these dispute causes is easily resolved. They either did or did not happen. The fourth item is trickier, and the owner will usually call on the architect to defend why he considers the work to be unsatisfactory and who is responsible for correcting it. With finish items, where the quality concern is not seriously detrimental to the owner, and the contractor is adamant in his position, the owner may overrule the architect and accept the work. That is his right, and happens more frequently than might be expected. Architects should accept this right of the owner as long as the work in question does not render any portion of the facility unsafe or, in the opinion of the architect, is not in compliance with the code. Architects should dutifully record the owner's acceptance of the work.

Punchlist inspections occur at the end of projects when contractors are asking for release of retainage and much of the schedule of values has been invoiced. Owners often argue that they need to hold retainage to compel the contractor to complete the punchlist. On small projects with limited retainage and a large number of punchlist items, this may well be the case. On larger projects, the amount of money necessary for the owner to hold should be a subject of discussion among owner, architect, and contractor. Public work usually has strict requirements regarding retainage, but private work allows the owner and contractor to agree on a fair percentage of project funds to hold. The owner's interest should be adequately served by holding a sum that is comfortably large enough to ensure that, if the general contractor disappears from the project, the owner can hire others to complete the punchlist items. This is, like so many aspects of construction administration, more art than science. In general, however, unless there is a substantial quality issue that concerns the owner, holding all the retainage while the punchlist is being completed is considered by some to be unnecessarily punitive. The contractor is usually a cooperative partner in this venture. The vast majority of the punchlist work is being performed by his subcontractors, so he wants to hold a healthy sum from them as well to ensure they will respond. He is more than willing to use the owner as the heavy in this endeavor, but he also recognizes that returning some portion of the retainage for work completed and accepted by the architect is fair. The architect and owner should be guided in this area by their experiences with the contractor on the current and previous projects.

SCHEDULE CLAIMS

Contractors often display a Jekyll-and-Hyde approach to schedule claims. Some, particularly those operating under a liquidated damages clause in their contract, will carefully monitor schedule delays and alert the owner when he has been delayed by the code officials, architect, weather, other prime contractors, or the owner himself. In relation to the architect, RFI and submittal logs will show the contract required response time (or the contractor's assumption of a reasonable time), and how many days the architect was ahead or behind that time. However, a simple tally of these plus and minus changes to the schedule may not be appropriate either to support delay claims or to defend against them. Change order requests will include an addition to the construction time allotment with each request, often ignoring the fact that the change order work has no impact on the overall schedule. These tactics are essentially defensive in nature, intended to protect the contractor from a large liquidated damages dispute with the owner at the end of the project. Sometimes they can be a prelude to a claim, especially where the contractor's bid was substantially below that of the next lowest bidder, or where he is basing a claim on perceived impacts of a delay to his schedule. This claim will appear at the extreme end of the project immediately before final payment, and is usually accompanied by a new critical-path schedule and list of delay causes.

A contractor has a right to expect reasonably prompt responses from the owner and architect on submittals, RFIs, color selections, and other requests he makes of them. Where the architect has agreed to a submittal and RFI response time in his agreement with the owner, or consented to its inclusion in the owner-contractor agreement, he should abide by this agreement. Real delays, and genuine general conditions costs, can occur on projects when the design professionals do not respond in a timely manner to issues affecting the schedule.

That said, some contractors will treat any missed deadline as a delay in the project, and therefore worthy of some compensation. In a common example, they will accrue the architect's delays in responding to RFIs and submittals, add in extra days for the change orders, and treat the sum as a bulk extension of the construction time—to which they will apply their daily general conditions rate. This ignores the *critical path method*—or the planned sequence of events of a project—that they employed in planning the work, and also does not address any missed opportunities the contractor could have used to advance the work during the course of the project. Contractors employing this technique also do not usually offer the owner the credit of time saved through quick responses, faster payments, unusually beneficial weather, or other advantages. They will gradually tally as many of these small delays as possible in the hope that the total claim will yield some settlement from the owner at the end of the project.

Schedule claims can become very complex, very quickly. Numerous construction consulting firms exist to either prepare schedule claims on behalf of the contractor, or to defend the owner against them. The difficulty in reviewing claims based on a disruption in the schedule is that the claim often argues that the critical path was disrupted by a change order, a delay of some type, or the failure of the owner or his design professionals to respond quickly enough to an issue. The contractor making a critical path claim is saying that a portion of the project was held up, and that in turn delayed all other events that relied on its completion (see Figure 8-1).

DEFENSES AGAINST CLAIMS

When a dispute between an owner and contractor defies negotiations and all normal attempts at resolution, the attorneys will take over and guide the owner through mediation, arbitration, or litigation, as described in Chapter 7. The architect will become an advisor to the owner and impartial arbiter of the construction docu-

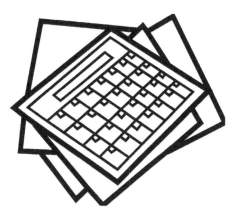

- Weather
- RFI and Submittal Delays
- Product Delays
- Subcontractor Staffing or Scheduling Issues
- Change Order Delays
- Payment Delays

FIGURE 8-1
Types of contractor schedule delay claims.

ments during this period. The owner and his attorney may employ a variety of defenses against the contractor's claim. It is worthwhile for the architect to be aware of the most common defenses used by owners to defeat contractor claims.

Notice Requirements

The construction documents, or the contract for construction, may state time periods in which change order or schedule claims must be made. In those cases where contractors make claims at the end of a project, well after work has been covered and the owner has a reduced ability to research the costs or seek less expensive alternatives, the owner can claim that the contractor failed to provide timely notice of his claim.

Prior Settlement of an Issue

Where an owner has previously agreed to settle a contractor's claim for additional compensation or time, the contractor cannot usually return to the owner for additional compensation later. This assumes, of course, that circumstances have not changed regarding the original issue. A contractor's claim for a schedule delay relating to a change order is such an example. If the contractor did not claim additional time when the change order was negotiated and signed, the owner can assert that the contractor is precluded from claiming the additional time later in the project.

Exculpatory Clauses

An exculpatory clause is a statement included in the contract that limits the owner's liability for schedule or other claims by the contractor. A common example is the "no damages for delay" clause often used by owners and architects to shift responsibility to the contractor for managing the schedule of a project. While an owner agrees to pay extra money for extra work, or even for extra time required of the contractor, as a way to avoid a fight over liquidated damages, this clause effectively tells the contractor that no money will be paid for delays. Additionally, each approved change order should state that the change order amount includes compensation for all extra time resulting from the change order work and related work, and noting that the contractor may not claim additional delay or general conditions expenses related to the work covered in the change order. Enforcement of these clauses varies with the jurisdiction, but their reasonable use is generally held to be valid as long as it is not contrary to public policy.

Breach of Contract

Where a contractor is making a claim against an owner under a contract that has not been fully or properly performed by the contractor, the owner may be able to argue that the contractor is in breach of his contract, and therefore not entitled to his claim or full payment under the contract.

Abandonment

A contractor who abandons his obligations under the contract may be liable to the owner for damages, or may be unable to collect additional claims he makes against the owner. A contractor, for instance, who abandons his punchlist completion and warranty obligations because of a dispute with the owner on unrelated items may find his claim undermined by the owner's counterclaim that he abandoned his responsibilities under the contract.

Excusable Delay

If an owner has not assumed the risk of a certain event, the owner may not be liable for extra costs incurred by the contractor as a result of the event. As a result, a contractor may be entitled to a time extension but not be entitled to additional compensation from the owner for extra costs resulting from the event for which the owner did not assume the risk. For example, a contractor may be entitled to a schedule extension for delays caused by excessive adverse weather conditions but may not be entitled to recover additional costs arising from the delays if the owner did not explicitly assume the risk of adverse weather in the agreement he signed with the contractor. A court may find that adverse weather is an inescapable and predictable aspect of the construction industry, and that a contractor should anticipate a normal pattern of inclement weather in bidding and scheduling the work.

Statutes of Limitations

A statute of limitations limits the amount of time a party can take to make a claim against another party for damages. This limit varies by jurisdiction and action, and there is even some interpretation possible in construction issues as to when the damage occurs. Still, it is possible for the owner to use this defense if the contractor's claim occurred far enough in the past to trigger a statute of limitations question. The architect should include in the supplementary conditions a particularly short period of time after substantial completion during which the contractor can submit claims, and remind him at substantial completion that this restriction exists and *time is of the essence.*

Liability of Others

Sometimes, a claim against an owner may occur where the owner has no fault. Where the claim occurs as a result of the actions (or inaction) of an agent or representative of the owner, the owner may be liable for the claim. In other situations, however, the owner's contractual relationship with a third party may relieve the owner of liability for the errors of the third party, and even for a contractor's mistaken reliance on their authority to act on behalf of the owner. An example would be where the owner relies on an environmental testing company to perform testing for environmental hazards and advise the owner and his professional team how to mitigate the discovered hazards. If the environmental testing company issues instructions directly to the contractor without the owner's approval, and those instructions cause additional costs to the project, the owner may hold the contractor responsible for improperly act- ing on the instructions. The owner's contract, for instance, should expressly disavow any financial responsibility for other contractors working for him or others.

Mitigation

The contractor is obligated to mitigate any damage that results from a problem on a construction project, whether his fault or the fault or others. When a contractor makes a claim to correct damages that could have been substantially reduced by some reasonable action that he failed to take (such as notifying the owner sooner), then the owner may have a claim against the contractor for the excess damages. An example would be the bursting of a water pipe (not the fault of the contractor) on the upper level of a facility. If the contractor does not take reasonable steps to protect finished construction below the place where the problem occurred, and additional damage occurs as a result, the owner may seek to make those damages the responsibility of the contractor.

Misrepresentation

When it is clear that the contractor has presented an owner with a claim for extra work that was never performed, is grossly inflated in cost, or is supported by documents that are not valid, the owner may have an action against the contractor for fraud, or intention- ally deceitful conduct. This is a rare and extreme occurrence. While inflated change orders are common, they more often represent a subcontractor's or contractor's desire for excess profit more so than outright fraud. The architect should advise the owner and the owner's attorney when he suspects fraud.

RETAINAGE

Retainage is a useful tool in protecting the owner from defects in the work that are not immediately evident, or in ensuring that the contractor performs punchlist work in a timely and acceptable manner (see the Risk Hazard Flags). Retainage has been a part of construction contracts for more than a century, and for good reason. Owners value retainage because of the leverage it provides them, particularly at the end of a contract.

Contractor Incentive

Holding retainage gives the contractor a financial incentive to stay on the job, work until completion, complete the punchlist, and correct any defects. The retainage constitutes a small percentage of the payments made to the contractor on each payment application, but by the end of the job the total amount provides a strong economic incentive to complete the contract for construction.

 **Risk Hazard Flags**

Retainage Liability Dangers

- Reducing retainage without owner approval
- Holding full retainage past substantial completion
- Holding retainage for the one-year warranty period
- Returning all retainage prior to completion of punch list
- Returning all retainage prior to receiving closeout documents
- Retainage still owed by contractor to subcontractors after release of retainage by the owner

Funds to Remedy Defects

If a general contractor fails to complete some portion of a project, or completes it poorly, the retainage provides an immediate source of funds for the owner to use in correcting the problem, particularly when it occurs near the end of a project. Subcontractors and suppliers also benefit from retainage because it can be used to pay them if the contractor defaults.

Contractors on large projects often clamor or negotiate for a reduction in retainage to 5 percent when the project reaches the later stages (approximately at the halfway point of the project). Although this is often argued as being "standard procedure" on projects, absent any mention in the agreement, it is a pure courtesy extended by the owner for the benefit of the contractor. Where 5 percent retainage will, in the estimation of the architect, reasonably cover defects in the remainder of the work and the performance of the contractor to date has been satisfactory, the architect may recommend to the owner to grant this request. In practice, no retainage is held from any application for payment after the point of agreement, so the percentage of retainage slowly declines to approximately 5 percent by the end of the project. There is no single large payment made to return half the retainage when the owner agrees to the reduction. This is sometimes confusing to contractors, who expect an immediate return of half the retainage held up to that point.

A conflict point with retainage often occurs at the end of the project, when the owner has beneficial use of the facility but has not agreed to final acceptance due to outstanding issues with the punchlist, change orders, or other minor matters. The contractor may demand return of the retainage, arguing that he is owed it and that the act of the owner in holding the large sum of the retainage is a punitive means of forcing the contractor to settle lesser issues on the owner's terms. See the C.A. Anecdote for an example of a real-life problem relating to retainage.

When the project is on the verge of final acceptance, closeout documents are complete, the contractor has otherwise performed well, and there are no exceptional issues that represent a financial threat to the owner, the architect may advise the owner to return all retainage except a reasonable amount that will safely allow the owner to complete all the remaining punchlist items should the contractor fail to do so. Full retainage may be held by the owner for a number of legitimate reasons, however, including the following:

- Where genuine concerns exist regarding latent defects in the work, missing final releases of liens, or the contractor's veracity in his certification that he has paid all debts and claims

- For exceptional and worrisome punchlist items, the architect should advise the owner to keep all retainage as both an inducement to the contractor to complete his obligations and as a financial safeguard in the event the contractor defaults late in the project.

Some owners erroneously view retainage as money that can be held for the entire warranty period, as a way to ensure that the contractor meets his obligations in the year following final acceptance of the work. This cannot be done unless it is expressly included in the agreement between the owner and contractor.

Other owners will insist on holding all of the retainage until every punchlist item is complete, all closeout documents are delivered, and they have satisfied themselves that all the building systems are working properly.

The decisions of when to reduce or return retainage are often made on a project-by-project basis. Much depends on the architect's and owner's comfort level with the performance of the contractor, and the relative risk associated with returning held money too soon.

Rules of Retainage

General practice over a long period has created the following commonly accepted guidelines for managing retainage:

- Retainage should be 10 percent of the total value of a contract, but in no case should it fall below 5 percent until the point of when a certificate of substantial completion is issued by the architect.

C.A. Anecdote

The Retainage Conundrum

The Problem

Joe's client was on the phone, and didn't sound happy.

"I just got the latest application for payment. You're recommending I give away half my retainage. That doesn't make any sense."

"I understand your concern," said Joe. "But Plumb Bob Construction has finished most of the punch list and you're already occupying the facility. He has delivered the operations & maintenance manuals, but still owes us the as-built drawings. I think the remaining retainage is adequate to protect your interests."

"I don't agree," said his client. "I don't see why I should have to give any of it back until the one-year warranty period is over. If I return his retainage now, I don't have any way to force him to fix anything that goes wrong during the next year."

"That isn't the purpose of retainage. Retainage should be returned to the contractor with final payment when he has met all the obligations of his agreement. He's close to that point now, and returning 50 percent of the held retainage leaves more than enough to compel him to complete the last few punch-list items and deliver the as-built drawings."

"Yeah, and when I call him in three months after my water heater bursts, all I'll hear is him laughing before he hangs up."

"Let me ask you a question," said Joe. "You sell appliances. When someone buys a washer from you, pays for it in full, and it quits working a few months later, who do they call?"

"You know darn well they call me. We service everything we sell, and honor our warranties."

"And you do this because you've been in business for years and your reputation is important to you in sustaining your business. We chose Plumb Bob Construction for the same reason. They have been in business for twelve years, have excellent references, and there is no reason to think they will not honor the certificate of warranty they have already provided to you with their close-out documents."

The Resolution

Joe's client still felt uneasy about giving back all the retainage. Plumb Bob Construction offered to pay half the cost of a maintenance bond, which provided Joe's client with the extra assurance he was looking for during the warranty period that he had some recourse if any warranty problems were not handled by the contractor. With the bond in hand, the client agreed to return all retainage to Plumb Bob with the final application for payment.

- Retainage should be equitable. General contractors should not withhold more retainage from subcontractors than the owner holds from them.

- Contractors should pay subcontractors their retainage when the owner releases retainage to the general contractor for the subcontractor's work.

- Owners should release most retainage upon substantial completion, holding only enough (a value at least equal to the amount of work remaining) to ensure completion of the punchlist and closeout documents.

- Owners should release all retainage with the final payment, unless stipulated otherwise in the contract.

LIENS AND RELEASE OF LIENS

A construction lien (also known as a mechanic's lien, laborer's lien, materialman's lien, or supplier's lien) is a form of security interest granted over the title of a property to secure the payment of a debt or claim. Liens may be filed by contractors against an owner, or by subcontractors and suppliers against the owner, to force payment for work they have performed but for which they have not been paid. Liens can even be filed by architects against the owner in certain circumstances, and within some time limits in various states. A lien prevents the owner from borrowing against the value of his property, selling it, or transferring the property to another owner until the debt is paid. The party who files the lien is referred to as the "liener," while the victim of the lien is referred to as the "lienee."

The most common cause of liens in construction projects is the failure of the contractor to pay a subcontractor or supplier. In fewer instances, the general contractor will lien the project to protect his interests when he fears the owner will not pay him for claims he has made. Although the owner is the lienee in each case, the lien is actually filed against the project itself—literally the value of the constructed work.

Liens came into existence because contractors often find the need to protect themselves from owners who do not pay the bills for the goods and services that comprise the contractor's construction effort. While one strategy for the contractor is to avoid allowing the owner to get too far ahead of him by withholding payment for accepted work, another form of protection is the filing of a lien. A properly and legally filed lien acts as valid legal notice to all the world (but especially to potential purchasers) that a party providing goods or services for the project has not been paid. Therefore, any transaction to convey or transfer the title of the property, such as a sale to a third party or the granting of a mortgage to a bank, is encumbered by the debt. Once a valid lien is filed, title insurance companies will not insure the property title, and may only permit a sale or transfer of the property to move forward if the money associated with the lien is set aside from the sale proceeds (sometimes said to be held in escrow).

The design professional must be familiar with the lien process because he may be called upon by his contract with the owner to arbitrate the amount, or even the existence of, a valid lien. Even if the architect is not the arbiter of the lien, he may nevertheless be required to testify in court regarding the work which is the subject of the lien. Since a lien creates specific rights in the property owned by another, most states have statutes describing in great detail the procedure required to file a lien. These statutes typically require a contractor to certify under oath that a specific amount has been properly invoiced under the agreement between the property owner and the contractor, that the contractor has demanded payment, and that the lienee has wrongfully refused to make the payment. There are often very strict rules as to where and when the liens can be filed. The statutes for liens, sometimes referred to as mechanic's liens, often require mini-hearings to be conducted by professional arbitrators (often sponsored by the American Arbitration Association) to determine the validity of the lien. There are also statutory penalties for persons who wrongfully attempt to file a lien, such as when there is no contract for the goods or services or where there is a legitimate disagreement over the price, quality, or quantity of the goods or services. Improper filing of a lien against a property can still interfere with its sale, financing, and the free transfer of property title. Properly used, the lien can be a powerful tool to require owners to pay their debts in a timely fashion and take responsibility for the financial risks associated with construction.

The architect can also become embroiled with the lien process, however, when the contractor (rather than the owner) is the party who fails to pay construction debts. Subcontractors of various tiers, vendors, and material suppliers usually have the right to file liens against the property on which they have provided construction goods and services. The purpose of their lien is to serve notice to the owner, and others, that the contractor is not paying his debts. The owner then has the opportunity to demand an explanation from the contractor for why he is not paying his debts. If the owner does not receive a satisfactory explanation from the contractor, he can withhold payment from him to satisfy the debt, or increase the retainage he is holding to protect against such problems. In most situations, the owner has the right to pay the liener directly to wipe out the lien and avoid its onerous effects.

The owner whose property is affected by liens, or the subcontractors who filed the lien to force payment of his debt, may each attempt to hold the architect responsible if he approved payment requests from the contractor when he knew, or should have known, that the contractor was wrongfully withholding payment from his sub. The owner and subcontractor could argue that the architect's role in approving payments jeopardized the owner's interest in his property, or the ability of the subs and vendors to obtain payment.

There are various strategies that the architect acting as contract administrator can employ to limit his liability in these situations and from liens in general, including:

1. The architect's contract with the owner, and the front-end section of the contract documents, should expressly exclude any responsibility on the part of the architect to investigate for or take action regarding liens asserted by any person providing goods or services for the construction.

2. This exclusion should be repeated in all forms used by the architect to document the review and approval of payment applications by the contractor.

3. To make this exclusion applicable to all subcontractors and vendors, the architect needs to ensure that all persons in contract with any of the contractors or the owner be obligated for this part of the agreement as if they had executed it directly. One easy way to do this is to simply require the contractor to incorporate, by reference, all of the requirements of his contract with the owner (but not the owner's payment requirements to the contractor), in any contract or purchase order for goods or services the contractor executes with his subcontractors or suppliers.

4. Requiring each of those parties to do the same with all sub-subcontractors and vendors, of whatever tier, provides substantial additional protection to the architect.

5. The contractor should also be required to indemnify the owner and the architect for any lawsuits, including payment of legal fees for claims brought by subcontractors or vendors through liens.

6. The architect can also require the contractor to obtain partial releases of liens from all subcontractors to be attached to each application for payment during the course of the project. At the completion of the work, it is especially important to obtain a final release of lien at the time of final payment.

Most states have rules invalidating waivers of liens by contractors or subcontractors, but the release of liens provides a measure of notice to the owner and the architect if the subcontractors or vendors have valid lien claims. The refusal of a subcontractor to provide a lien release is a powerful signal that payment issues exist on the job site.

This all seems difficult enough, until one considers additional complications that can arise from the filing of a lien, such as:

- The contractor may consider the subcontractor who liened the owner to be in default, to have performed the work poorly, or to otherwise not be entitled to the amount he has claimed.

- Similarly, the owner may consider the general contractor to be in default and feel that he has withheld payment for good cause.

For the pain they can cause, liens have one overriding value to the architect in his services to the owner: They focus attention on a problem. If a contractor has sufficiently alarmed a sub or supplier to the point where they feel compelled to lien the property, there is almost always a deeper financial problem. A quick investigation by the architect or owner may find that there are other subs and suppliers who are complaining that the contractor is late in paying them for work already paid for by the owner. In other instances, the filing of a lien may reveal a deep concern on the part of a contractor with the performance of a particular subcontractor. Like the owner, the contractor views holding money as the main motivator of good performance from his subcontractors. Where a sub has not met the schedule, workforce, or quality obligations he owes the contractor, withholding funds is the way in which the general compels better performance from his sub. Identifying the true cause behind a lien filing is often the responsibility of the architect. Once he is confident that he understands the foundation of the lien, he can advise the owner on what, if anything, he should do in response to the lien filing.

As long as a contractor's failure to pay a sub is not part of a deeper financial concern, the owner may rightfully say to the contractor, "Take care of the problem and provide me with a lien release."

Neither the owner nor the architect should interfere in the contractor's right to enforce performance from a subcontractor, but neither can they ignore the fact that a lien filing can be a harbinger of serious financial problems for a contractor.

It is critical for the architect to identify all principal subcontractors and suppliers during the course of the project, and as noted, to ensure that partial lien releases are obtained with each periodic application for payment, and that final releases are obtained at the conclusion of the project. On smaller projects, the owner may forego periodic releases in the interest of reducing paperwork, though this will prevent him from learning of payment issues as early as he otherwise would. The closeout documents typically require both a final release of liens from the contractor and principal subcontractors, as well as an affidavit of payment of debts and claims by the contractor. This affidavit provides the owner with certification by the contractor that he has paid all known monetary claims against the project. While it does not ensure against the filing of a lien by an unknown subcontractor or supplier, it does provide the owner with a legal basis for action against the contractor to settle the lien.

As a postscript, the architect should read carefully all lien laws in the states in which he operates as many states allow providers of professional services such as architects, engineers, and surveyors to assert liens for unpaid invoices in the same manner as the contractor or subcontractor.

CLOSEOUT DOCUMENTS

Closeout of a project involves a wide range of actions by the contractor in administratively and physically transferring the facility to control of the owner. Perhaps the most important of these actions is the administrative closeout of the contract through the submission of the closeout documents (see Figure 8-2). Here is a list of standard closeout documents an architect should expect to receive from a contractor:

1. An application for payment showing all work as completed and requesting return of all retainage

2. A punchlist showing all items identified by the contractor and architect, certified by the contractor as being completed

3. A properly executed consent of surety for final payment (if the project is bonded)

4. A final release of liens for the total amount of the contract sum

5. The as-built drawings

6. Warranty and maintenance information

7. A properly executed affidavit of payment of debts and claims

8. Originals issues of the certificate of occupancy (if not previously provided to the owner).

Review of the contractor closeout documents is a time when the architect is called upon to be particularly careful about the completeness and appropriateness of the documents he is reviewing. If he has any concerns regarding whether key legal documents are

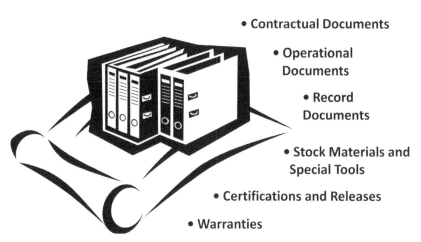

• Contractual Documents

• Operational Documents

• Record Documents

• Stock Materials and Special Tools

• Certifications and Releases

• Warranties

FIGURE 8-2
Closeout documents required from the contractor.

correct, he should ask the owner to make his attorney available to review them prior to approval. Once the final payment is made, there is little opportunity or incentive on the contractor's part to revise and resubmit a document that is deemed incorrect. The time to require revisions is when the final application for payment is pending, when money is being held.

OPERATION AND MAINTENANCE MANUALS

At the end of a project, the contractor owes to the owner an organized set of operation and maintenance (O&M) information for all major equipment installed in his facility. Typical specification sections call for this to be provided in three-ring binders, organized by trade, and prepared in duplicate. A normal specification section for this requirement will read:

Assemble a complete set of operation and maintenance data indicating the operation and maintenance of each system, subsystem, and piece of equipment not part of a system. Include operation and maintenance data required in individual specification sections and as follows:

1. ***Operation data***
   a. Emergency instructions and procedures
   b. System, subsystem, and equipment descriptions, including operating standards
   c. Operating procedures, including startup, shutdown, seasonal, and weekend operations
   d. Description of controls and sequence of operations
   e. Piping diagrams
   f. HVAC balancing report (if applicable)
   g. Commissioning report (if commissioning services are required).
2. ***Maintenance data***
   a. Manufacturer's information, including list of spare parts

b. Name, address, and telephone number of installer or supplier

c. Maintenance procedures

d. Maintenance and service schedules for preventive and routine maintenance

e. Maintenance record forms

f. Sources of spare parts and maintenance materials

g. Copies of maintenance service agreements

h. Copies of warranties and bonds.

It is beneficial for the architect to involve his engineer in reviewing this information, since the mechanical, electrical, and plumbing engineer was likely responsible for specifying the equipment covered by this requirement, and perhaps even for writing the requirement itself. It is perhaps more helpful for the architect to pass the O&M to the owner's facility manager, plant engineer, or other party responsible for facility maintenance. They may have specific questions or concerns regarding the O&M information, and the time to obtain the answers is when the contractor and his subs are seeking final payment and return of held retainage.

OWNER SYSTEMS TRAINING

Even owners with basic monitoring, management, or physical systems need to know how to operate them. Specifications should include provisions requiring the contractor to provide basic training in how to operate equipment installed in the facility to the owner's facility personnel prior to closeout. For skilled facility managers, this instruction is usually limited to those pieces of equipment that they are unfamiliar with or that represent custom installations with procedures that are unique to the installation. In general, the architect should consult with both the owner and contractor at the time of closeout to determine whether the owner is satisfied that his people are familiar with the operation of all equipment installed under the contract for construction. At the least, the architect should ensure that the owner has received training in the operation of pressurized systems of any type and life-safety systems, including fire alarm and fire protection (sprinkler) systems, and security systems.

Typical specification requirements for owner systems training are as follows:

1. Provide instructors experienced in operation and maintenance procedures.

2. Provide instruction at mutually agreed-on times. For equipment that requires seasonal operation, provide similar instruction at the start of each season.

3. Schedule training with the owner with at least seven days advance notice to the owner and architect.

4. Coordinate instructors, including providing notification of dates, times, length of instruction, and course content.

5. Develop an instructional program that includes individual training modules for each system and equipment not part of a system, as required by individual specification sections. For each training module, develop a learning objective and teaching outline. Include instruction for the following:

   a. System design and operational philosophy

   b. Review of documentation

   c. Operations

   d. Adjustments

   e. Troubleshooting

   f. Maintenance

   g. Repair.

AS-BUILT DOCUMENTS

All contractors should maintain as-built information as the work progresses. Contractors handle this in different ways. Some place the burden on each sub to keep records for his trade independently and turn it over to him at the end of the job for transfer to a master set. Others require the foreman for each subcontractor to record the information on a master set kept in the field office. Still others forget about it entirely and can offer little to the owner at closeout time.

Since this is still largely a paper-driven enterprise by contractors, some owners pay the architect to take the contractor's field set and transfer the as-built information to electronic files so they have a more readily-usable record document. In most private contracts, however, the contractor remains solely responsible for maintaining the as-built set and turning it over to the architect for review at the end of the project. The architect's role is that of both a facilitator and a nag.

- Facilitator: If the contractor requests electronic files or a physical hard copy for the purpose of keeping as-built drawings, the architect should accommodate him.

- Nag: At every project meeting, particularly during the period when underground work is occurring, the architect should ask to see the as-built drawings and remind all of their importance to the owner (and the fact that they are a requirement of the contract).

FINAL PAYMENT

The final payment request from the contractor represents the last payment from the owner to the contractor. That part is clear enough. Where the difficulty often occurs is when the owner and architect are not satisfied that all conditions of the contract for construction have been met (see the Risk Reduction Tools).

Final payment will normally consist of the return of the remainder of retainage. Usually it is not the full value of retainage since that sum should have been reduced at substantial completion to an amount that the owner and contractor agree fairly represents the value of the remaining work. The retainage can represent a substantial portion of the contractor's profit for a project. Therefore, the retainage is excellent insurance for the owner that the contractor completes the punchlist work, properly executes all warranties and

Risk Reduction Tools

Final Payment Checklist

- Certificate of occupancy issued by code official
- Affidavit of payment of debts and claims received
- Final release of liens from contractor and subcontractors
- All closeout documents received and accepted
- All punchlist items completed
- Owner in agreement with final payment
- Final application for payment received
- Certificate of substantial completion, signed by all parties
- Certificate of occupancy issued by the municipality

close-out documents, and assembles all required record drawings and specifications per the contract documents. All punchlist items should be complete before final payment. All open change orders, schedule claims, or other disputes should also be settled before final payment. While the owner should not use a delay or withholding of final payment as a punitive device, he may certainly ask the architect to remind the contractor that final payment will not be made until all close-out documents are received.

It is not appropriate for owners to hold retainage through the end of the warranty period. Owners will sometimes argue this is necessary as a way to ensure a response from the contractor in the event a warranty claim arises. The General Conditions of Construction should specify that retainage will be returned at the time of final payment. An owner is sufficiently protected during the warranty period by the contractor's warranty guarantee, a legal document that obligates him to respond to owner requests during the warranty period.

Another reason that owners decline to make final payment is because they may consider some aspect of the work to be defective or poorly installed. An owner, for instance, may have legitimate concerns regarding cracks that are appearing in the concrete slab of a warehouse. Are they surface or shrinkage cracks that can be easily repaired, or are they indicative of a larger problem that may render the slab unusable by the owner? In these cases, the owner rightly looks to his professional team to offer advice on how to resolve the situation. More obdurate owners may insist that the entire slab be removed and replaced, arguing that anything less means they have been denied full value of their contract. The architect, willing or not, may be the best person to resolve the issue fairly. He can utilize his engineer to review the situation and render an opinion, offer suggestions for testing to determine if the cracks are structural or superficial, and recommend a repair procedure for the latter. Because of the architect's special relationship with the owner, one that preceded and is often stronger than that between the owner and contractor, the architect is uniquely positioned to counsel the owner on fair and measured responses to legitimate quality concerns. As with any other construction dispute, the architect should endeavor to keep the situation fluid through helping each side steer clear of hardened positions that do not leave any room for a reasonable compromise. Construction quality concerns always loom larger at the end of a project, when an owner feels that he should be receiving perfect work for the monies he paid. It is difficult for architects to explain to owners the sometimes unpredictable nature of building materials, and the fact that an otherwise satisfactory installation may yield unsatisfactory results. Problems that occur through no fault of any party may be the responsibility of the owner, and that is a hard thing for any owner to accept. As difficult as this is, helping the owner and contractor find their way through a final payment dispute to a satisfactory conclusion is one of the highest services the architect can perform—and ultimately in his best interest as well.

9

After Construction Ends

WARRANTY WORK

Long after the project is over, the architect can receive a phone call from his client. This call usually occurs after a heavy windstorm, rainstorm, or hard freeze, and consists of the news that something is wrong with the building. It is perhaps heartening that the client would think of the architect as the first person to share this news with, but it is hardly a welcome call. If the call occurs within a year of substantial completion, the problem may be the responsibility of the contractor to correct under the standard one-year warranty required in most states, and under all contracts that use standard documents offered by the American Institute of Architects.

It is not necessarily the case that any problem occurring in the first year after substantial completion is the responsibility of the contractor. The contractor may investigate the problem and determine that the owner made modifications or failed to perform routine maintenance that caused the problem. Nevertheless, the architect's first phone call is to the contractor, asking him to check out the problem and correct it if it is within the scope of the warranty.

It did not have to be this way. One of the required closeout documents often missed is a list of subcontractors responsible for the various aspects of the work, as well as their contact information. The owner then has a direct means of contacting the appropriate sub in an emergency situation, such as when roofing is blown off the building. Although he should always contact the contractor if possible, warranty work is a situation where the owner may feel free to contact a sub directly if he needs immediate assistance. It is important to remember that the warranty to the owner is from the contractor—not his subcontractors. So if a particular subcontractor goes out of business or simply fails to perform warranty work, the obligation falls back to the contractor to perform the work himself. In situations where life or property is at stake, and neither the contractor nor his sub respond to the owner's request for assistance, it is the owner's responsibility to minimize his damages and protect his interests by arranging for emergency repairs to be performed. He may deal at a later date with the contractor's failure to perform. He cannot allow the problem to deteriorate unreasonably and thereby manufacture a larger claim against the contractor.

The architect may be obligated under his agreement to assist the owner in preparing a list of warranty items at a certain date after substantial completion. By that time, the owner should have tested every new installation in the building and have a reasonable idea of what requires repair by the contractor.

Occasionally, a contractor or his sub will perform warranty work for an owner within the one-year period, and later attempt to bill the owner for this work with the claim that the architect's detailing was flawed. This claim of a *design defect* often involves some aspect of roof flashing, water management, or mechanical systems, and argues that since the installation was performed exactly in accordance with the construction documents, the problem must lie elsewhere (the architect's details). The architect may claim in his defense that he did not design a roof to leak, and the problem is more likely a *latent defect*, a hidden flaw in the installation by the subcontractor that revealed itself over time. If, in this instance, the water infiltration is a one-time occurrence after a severe storm, it is quite possible that neither party is truly at fault. Severe storms test the best of installations, and a single occurrence during the warranty period may best be chalked up to the vagaries of construction. If, however, the leaks occur in several areas, or become part of a recurring pattern, then the owner will likely take more seriously the contractor's claims that the architect's flashing detail was perhaps not all that it should have been. The architect's best response is to participate in any meeting called by the owner to discuss the problem, if only to gather information regarding the likelihood of a design deficiency claim against him by the owner. Without admitting any fault, the architect may also offer to investigate the causes with the contractor. In roofing issues, especially, an alternative for the owner may be to ask a roofing consultant to investigate the problem and propose a solution. If the architect's detailing was sound, he should be vindicated in the process. If an independent roof inspection was performed at the time of substantial completion, the architect can produce this document as evidence that the roof was installed as specified. The architect should be careful, however, not to turn over the decision of fault to any consultant hired by the owner.

The owner benefits from other warranties offered by equipment and material suppliers. Since many of these warranty periods may run well beyond a contractor's standard one-year warranty, it is the owner's responsibility to pursue these warranty claims himself where the problem is unrelated to the contractor's installation.

AFTER THE WARRANTY PERIOD

The contractor may not respond to problems arising after the warranty period, depending on how he perceives the risk/reward value to checking out a complaint. The immediate question that may be difficult for the architect to answer: Is this a latent defect or the result of a failure of the owner to properly maintain his facility? If the problem is a repeat of a previously reported problem, or a continuation of a problem that was never fully resolved, the contractor has an obligation to address it. If the reported issue is a new issue that occurs several years after the end of the warranty period, there may be real validity to a contractor's response that he has no obligation to address the problem. This is especially true where the problem is the result of water detention over time or the failure of the owner to maintain the roof. The architect and contractor may choose to take a look at a complaint in the interest of preserving a relationship with a client who may give each of them future work. A secondary goal would be to satisfy their need to know about the actual nature of the problem at an early stage, and make a rough assessment of the liability associated with it. Most often, after-warranty calls are related to roofing issues, or other element-related infiltration problems. Sometimes the owner has failed to maintain his roof, or has employed other contractors who caused the problem. In other instances, a seam, coping, or flashing component has worked its way loose for any number of reasons and failed during a heavy storm. Outside the warranty period, this is an owner responsibility, which he may or may not accept. It is often difficult to convince an owner that he has a responsibility to perform regular roof inspections, maintenance, and repairs. Most tend to believe that some failure on the part of the contractor or his roofing sub resulted in damage years down the road. The owner may claim the work is a latent defect that resulted from improper installation and is just now revealing itself. He may demand an overall expert evaluation of the area in question, or insist on a repair and limited warranty from the contractor or his sub to protect against future issues. These demands will usually be declined by the contractor. Where the repair is simple and limited, the contractor may agree to perform after-warranty work in the interest of maintaining good relations with a client. The architect benefits from this action, of course, but no good deed goes unpunished. The owner may well expect the same response with the next roofing problem, and the contractor may find himself on call for the life of the roof.

RECORDS RETENTION

The unfortunate reality of architectural practice is that records need to be retained for as long as a project subjects the architect to liability. In a state where there is no statute of limitations or repose, this means forever. The purpose of retaining architectural files is threefold:

1. Preserving resources that will help defend the firm against a liability claim,
2. Preserving information that may benefit the firm in marketing or performing future work,
3. Retaining resources that may assist the client in repairs or emergencies.

While all of these goals are important, the benefit of retaining records to protect against liability claims is by far the most critical (see the Risk Reduction Tools). Many corporations advocate destroying e-mail and non-essential paper files once a project has been completed. Their legal staff has concluded that this is the best means of protecting the corporation from the liability of a "smoking gun" memo or errant e-mail that could have been created by any one of hundreds of personnel who may have been involved in a particular endeavor. The opposite is true in professional services liability. There is very little risk that any document a relatively small group of employees and consultants will create is likely to cause liability for an architect (absent, of course, an outright statement of admission to a negligent act or intent to violate a law or building code). Rather, the absence of documents, sketches, questions, concerns, and revisions will raise questions as to whether a design professional met the standards of professional care. An entire assemblage of the minutiae of architectural practice is powerful evidence that a design professional was doing his dead-level best to meet his client's needs and provide a reasonable standard of professional care. A blank slate puts the issue of care back into debate. The architect will argue he performed the necessary acts, while the client or contractor will argue he did not. The records could help swing the debate more in favor of the architect—if they exist.

 *Risk Reduction Tools*

Records to Retain when the Project Ends

- Submittals and shop drawings
- Original construction documents
- Architect's field set
- All project correspondence, including e-mails
- RFI responses
- Owner-architect agreement
- Additional services memoranda
- Project meeting minutes
- Punchlist inspections
- Project field reports
- Copies of permits
- Certificate of substantial completion
- Certitificate of occupancy

Minimum architectural records that should be retained:

- Submittals and shop drawings
- Construction documents, particularly the architect's field set
- All electronic files and e-mail communication
- All project correspondence, including e-mails and RFI responses
- Owner-architect agreement; additional services memoranda
- Project meeting minutes
- Submittals and shop drawings
- Punchlist inspections and project field reports.

Paper records should be retained at least until the statute of limitations expires in the state where the work was performed. In states where there is no statute of limitations for professional liability claims, the architect should consider scanning documents to eliminate the need for bulk storage of files.

Architectural practices should be extremely wary of emulating corporate practices commanding the destruction of normally archived information. Such practices may be considered "spoliation of evidence," done intentionally to thwart liability or a legitimate investigation. The person who destroyed the records, intentionally or unintentionally, may be subject to legal action as a result. Architectural practices should consult an experienced attorney to develop a regular records retention procedure—and stick to it.

CONSTRUCTION DISPUTES

Construction disputes are a fact of life. Often they originate with a disagreement over something that was unclear or misunderstood in the construction documents. Other times, they stem from the architect's rejection of a submittal or some part of the work. To some architects, disputes with a contractor represent a personal affront—an attack on their credibility as a professional. (See Figure 9-1 for typical types of construction disputes.) To others, they are simply a reality of the way we build, and must be accepted as a business aspect more than a professional assault. Most architects, though, probably suffer some regret that a contractor has found flaws in their documents or their handling of the project, and that the dispute may cost him prestige or cost his client change order money. He may realize that the drawings were not as thorough or well-coordinated as they could have been, that the contractor has some degree of right on his side, but that he must nonetheless do his best to defend his client and himself against the contractor's claims. A section in Chapter 5 explains how to handle adversarial relationships, and Chapter 7 includes techniques for responding to change order requests. This section's purpose is to discuss the general nature of disputes in construction, and to review the practical and legal aspects of managing them.

Disputes come from different sources, including:

- **Document-related disputes:** A detail is missing, unclear, or uncoordinated.

- **Rejection disputes:** The architect or one of his consultants has rejected a submittal, work-in-place, mock-up, or some other submission of the contractor.

- **Money disputes:** The architect has reduced the contractor's application for payment in areas where he judges the work has not been performed or was performed unsatisfactorily.

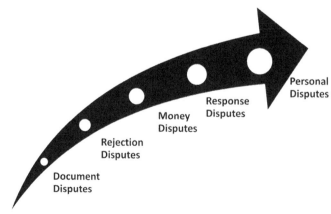

FIGURE 9-1
Types of construction disputes.

- ***Response disputes:*** The contractor does not believe the architect is responding in a diligent manner to RFIs, submittals, or pay applications. In many cases, this type of dispute originates from the contractor failing to get telephone answers to urgent construction issues.

- ***Personal disputes:*** Contractor or subcontractor personnel take personal offense at the comments or attitude of the architect or his consultants.

There are no easy answers on how to handle any of these types of disputes (see the Risk Hazard Flags). Architects who perform construction activities over a period of time develop their own techniques they are comfortable with for dealing with disputes, some with more success than others. Some architects enjoy verbally jousting with contractors and their subs, and take a certain delight in occasionally getting under their skin. Others are clearly uncomfortable on a construction site, and are wary of even speaking with anyone but the contractor's superintendent or project manager. It should be noted, however, that the architect who shies away from confrontation is likely not meant to be successful over the long term in architectural construction administration. Despite the differences in outlooks, there are a few basic truisms that seem to go with successful dispute management by architects, including:

1. ***Take action for a reason, and explain it.*** If the third mockup is still not acceptable, reject it, but state clearly why it is still not acceptable and what must be done to make it satisfactory. If the pay application for drywall exceeds the work in place, the architect should provide the contractor with his calculation showing why the claimed amount does not add up.

2. ***Be professional and courteous.*** Contractors and architects deal with each other from roughly equal positions of power. Each has his own role in the project, and each approaches it from different vantage points. There is no reason for an air of superiority on either's part. Each needs the cooperation and respect of the other and has a right to expect it. Personal disagreements and sniping between contractor and architect reflect poorly on both in the eyes of the owner. Even if other parties insist on personal attacks in their communications,

Risk Hazard Flags

Ways an Architect can Increase His Liability

- Architect does not respond to, or issue an interpretation in a dispute
- Architect is slow in responding to contractor RFIs
- Architect does not reference the documents
- Architect implies work is required beyond a reasonable extent shown in the documents
- Architect agrees with owner regardless of true feelings
- Architect is ignorant of owner/contractor agreement

the architect should avoid doing so, especially in print. This professionalism will pay dividends if the dispute ever reaches the point of litigation.

3. ***Discuss more than declare.*** If the contractor claimed 50 percent of the drywall was in place but the architect only counts 40 percent, discuss it together. The architect may find he missed a section; the contractor may find he over-counted. If the mason took offense to the architect's rejection of his work as "sloppy," the architect should talk with him about it. The architect may want to review the work again after it has been cleaned and rubbed. The mason may want to request a different mortar mix that is more workable.

4. ***Be ready to rumble.*** There are unreasonable and scheming people in the world. Construction probably does not have any more than any other profession, but it has its share. Any party who wants to challenge the architect's professionalism and skills needs to know the architect will defend his position and his client's rights under the contract. Conversely, architects are sometimes called upon to tell their client they are wrong and the contractor is right. This is painful duty, but sometimes necessary. Some truth-telling early in the construction process may save significant pain later.

MARKETING

Architects should be proud of their work. The completion of a building—any building—for an owner is an achievement. Not all buildings are destined for inclusion in an architect's portfolio, of course, but even if the building itself is not worthy of mention in a brochure, the client may be.

If the project ran beautifully, was finished on time, and the client is thrilled with his services, the architect should seize the moment and get photographs, a letter of recommendation, and references as soon as possible after closeout. If, like most projects, there were a few bumps in the road to completion, it might be better to let a little dust settle and give the client a few months before asking to take interior photographs. If the architect obtained the client's consent in the owner/architect agreement, it is a good time for the architect to gently remind the owner of this prior permission and schedule a date. Exterior photographs of facilities along public right-of-ways can be obtained easily enough without the client's consent, but engaging a professional photographer may require permission of the client and some degree of cooperation. This approach can be an opportunity to hold a "post-mortem" discussion with the client. There is a cathartic benefit to allowing clients to vent over what they felt were failings of the architect and their frustrations with the construction process. Without being overly defensive, it can be an opportunity for the architect to explain why he handled certain issues the way he did, and perhaps retain the client that might otherwise have looked for another architect for his next project. With a project that an architect wants to use in his portfolio to obtain other work, he should seek within a few months of completion to compile:

- Quality exterior and interior photographs (as appropriate)
- Photos of building details
- A record of basic project facts, including: area, final cost, construction time, completion date, and change order total
- Owner letter of recommendation, or agreement to serve as a favorable reference

10

Risk Management

Risk management means to take proactive steps to manage and limit risk. It is an essential fact of life in the realm of providing professional services, and is not far from every professional's mind as he deals with clients and contractors. Throughout this book, the topic has been front and center in virtually every issue addressed.

This chapter summarizes the major risks an architect will face in performing construction administration on any project. While certain projects pose special and unique risks because of their nature, there are a handful of problems that routinely clog the courts and arbitration hearing rooms. An architect who can eliminate as many of these common liability risks as possible will do much to prevent claims and litigation.

Here are the risks. Their successful management will aid the design professional in being more vigilant, and thereby improve his practice for his client's welfare and his own (see the Risk Hazard Flags).

TOP TEN CONSTRUCTION ADMINISTRATION RISKS TO AVOID

(The ten best ways for an architect to land in court)

Number 10: The Handshake Deal

(Working without a contract)

An architect is not an architect if he or she doesn't have any clients or projects. The temptation to take on work at any cost can

Risk Hazard Flags

Liability Warning Signs

Signs that liability trouble is brewing on a project:

- The owner is behind in paying the architect
- The owner is behind in paying the contractor
- The contractor is not paying subs or suppliers
- The contractor is claiming the architect is delaying the project
- The owner is claiming the architect is delaying the project
- The architect receives a large number of RFIs early in the project
- The contractor is late with major submittals
- The owner is refusing to acknowledge change orders
- A majority of change order requests are disputed by the owner
- The contractor is threatening to stop work
- The contractor replaces key subs in the middle of the project
- The owner receives a lien
- A worker is injured on the worksite
- There is little activity and few workers at the jobsite
- The contractor is falling behind schedule
- Owner's own forces are unresponsive or perform poorly
- Architect finds numerous quality issues during punchlist inspection
- Owner reports latent defects after occupying the facility

be great. It may even cause some architects to consider skipping a written agreement memorializing what the architect is expected to do and how much he will be paid for doing it, as well as other essential terms and conditions, for fear of scaring away the client. A client who balks at signing a legitimate price and scope agreement is a warning sign of impending trouble. It would be a mistake *not* to run away from such "opportunities." Each project should have a written agreement, signed by the owner, identifying as precisely as practicable the nature of the project, the architect's scope of work and, where appropriate, what it does not include. The peril to the architect and the open-ended problems that can result are obvious.

Risk Hazard Flags

Top Ten Hazards to Avoid

10. The handshake deal
9. Working for family or friends
8. Insufficient budget
7. Avoiding tough discussions with clients
6. Ignoring receivables
5. Ignoring submittal and RFI deadlines
4. Getting involved in means and methods
3. Not knowing your limitations
2. Not including a subrogation waiver clause
1. Working without insurance

Nearly as risky as having no contract at all is skimping on the agreement's other essential terms and conditions as previously discussed in this book, such as the roles and responsibilities of the owner, limitation of liability clauses, use and ownership of documents, procedures on termination or abandonment, and other critical issues. Many of these problems can be handled non-confrontationally by incorporating by reference a standard "terms and conditions" document. What an architect postpones making clear today will almost assuredly come back to haunt him tomorrow. With a range of available high-quality documents from the AIA and other sources, there is simply no reason for an architect to work without a written agreement of some type.

Number 9: Working for a Friend

(How to turn friends into foes)

Friends and family can be a fertile source of business for an architect. They offer friendly referrals and ready access to a wide network of potential clients. When a friend becomes a client, however, trouble may be brewing. The process of designing and building a project can be a harrowing and stressful experience for an owner, especially an inexperienced one. The owner needs to feel free to speak frankly to his architect about his fears, concerns, and complaints. Likewise, the architect should be able to exercise his professional judgment without concerns about personal feelings. Where the relationship is too close, it may be difficult for each of the parties to express himself openly and without reservation. The simple act of the architect charging appropriate fees, and the obligation of payments by the client, may create more anxiety than a standard friendship can bear. There is too much riding on the project for this to occur, and therefore an architect should think twice before working for a close friend or family member. If the architect does take the plunge in agreeing to such a relationship, he should set the ground rules early and stick to them. Make the rules be the bad guy and blame them frequently.

Number 8: Not Budgeting Enough for Construction Administration

(Who knew it would take so much time?)

Many architects calculate their construction administration fees according to the phases of architectural services. For example, they may budget 70 to 80 percent of the fee for design/procurement and leave 20 percent or less for construction administration, notwithstanding a full plate of activities for construction, which may take six to twelve months—or more. Often this is done to front-end load the fee and get most of the money in the architect's account by the time the "deliverables" are out the door. The problem with this approach is that it ignores the natural tendency to follow the money, and encourages a loss of interest in the project as the money runs out. This may result in a serious slip in services and the making of architectural decisions that are driven by fee rather than by professional responsibility or standards of professional care. Time should be taken to analyze the project scope and complexity, the desire versus the capability to provide contract administration services, and fair return for the services that need to be provided during construction. Some other factors to consider include the contractor's and owner's experience levels, the contractor's reputation for starting and finishing on time, and the likelihood that the job will be smothered in change orders.

Number 7: Avoiding Tough Discussions with the Client

(Not knowing when the honeymoon is over)

Often the relationship between the architect and owner, especially when new, is bright and shiny, like the future the parties contemplate for their project. Soon enough, however, the newness begins to fade, particularly as problems occur on the project. The architect who wishes to avoid bad feelings during this honeymoon period will shy away from confrontation of any type. Confronting the owner need not (indeed, should not) be adversarial. Early confrontation of problems usually leads to early solutions, which are typically less expensive and time-consuming than problems that are allowed to fester. The architect should get the client accustomed to the notion that there will be problems with construction (there almost always are!) and that a systematic approach of confronting, analyzing, deciding, and implementing is the best way to minimize the negative impact of construction problems. It is imperative for the architect to be frank, honest, and timely in tough discussions with his client.

Number 6: Ignoring Recurring or Unresolved Project Issues

(Someone should do something about this)

Often, the architect is called upon to meet with the contractor to look back at what has been accomplished to date and look forward in the schedule of future events. This is also a time to bring to the architect's attention problems or issues requiring clarification in the plans and specifications. In larger projects the meetings may be presided over by a construction manager who will make minutes of the meeting, often identifying issues and assigning a participant the role of resolving the issue or obtaining additional information. The contractor is usually very careful to record the results of the meetings in minutes that are dispersed to all the parties, with a request for a review and immediate response to correct errors. This role, especially where the architect is not the author of the minutes, is vitally important. First, the architect should review the minutes thoroughly for errors. They are more frequent than one might imagine, and are often detrimental to the parties who are not recording the minutes. Make sure that the problem/issue has been adequately identified, as well as the party actually charged with action and the level of action required. There must be a method to communicate any errors in the minutes that permits instantaneous notification of the error, as well as the correction required. E-mail is excellent for this purpose, as long as the sender insists on a response confirming receipt by all recipients (to ensure there are no glitches in transmission) and maintains a written copy of the transmission. Faxes showing the name and fax number of all recipients (and confirmation of a successful transmission) can also provide adequate proof of action toward correcting erroneous minutes.

Even after the architect determines that the minutes are accurate, he must make it a priority to take appropriate action and report it to the relevant parties. If the architect is responsible for providing a sketch, he should send it with a cover transmittal fax, advising the recipient about the purpose of the sketch and that it fulfills the obligation stated at the recent job meeting. Make sure that the item is covered at the next meeting and marked as completed. If the architect does not, rest assured someone will recall that he did not do what he was supposed to in a timely fashion and will assert that this failure was the cause of a delay claimed by the contractor. Maintaining a form for this purpose is an easy and quick way of assuring information concerning minutes is disseminated and not forgotten.

Number 5: Ignoring Deadlines on Submittals and RFIs

(You needed it when?)

Construction administration often means the architect is perpetually on call to answer questions about the design. Unfortunately, the contractor sometimes uses this as an opportunity to suggest that the design was deficient, and worse, the architect was complacent and inefficient about rectifying his own mistakes. It is not at all unusual for the contractor to keep track of the RFIs and submittals—including the request dates, demand dates (when he needs it returned), and the dates when they were actually answered by the architect. This is to document the project delays and place them squarely on the architect. For these reasons, it is essential to make RFI responses a priority without forgetting the principal role of the response—to clarify the design by telling the contractor what to build, while resisting the urge to tell the contractor *how* to build. The architect should keep a record or log of his communications for these purposes, with the dates he actually received and sent responses, and make requests of the contractor to confirm receipt of the responses. He should also set appropriate ground rules with the contractor's personnel regarding communication—and stick with the rules.

Number 4: Getting Involved with Means and Methods

(That's not in your job description)

Means and methods are the particular equipment, practices, and procedures a contractor uses to build a project. The knowledge of how to build is the job of the contractor. Architects may also know how to build, but they are not the contractor. An architect who wants to *act* as a contractor through offering advice on means and methods may end up explaining his theatrical talents in a courtroom. There are at least two ways that architects use poor judgment in dealing with the contractor's means and methods of construction. First, they often tell the contractor how to build, sometimes at the contractor's invitation. This happens most commonly where the architect suggests methods for erection of a building system or construction of a detail, or offers specific advice on remedying a deficiency in the construction, instead of demanding the contractor comply with the plans and specifications and holding the contractor to the contractual consequences when he fails to do so. The architect thinks he is expediting construction, when in reality he is interfering with the contractor's sole and exclusive province under the contract for construction. The contractor will be happy to take the architect's advice (he may even demand it) and then blame him for any deficiency that occurs—as well as the resultant delay.

Another way an architect uses poor means and methods judgment, ironically, is in his failure to comment in a timely fashion about issues which, although typically in the jurisdiction of the contractor, nevertheless impact the life, health, or safety of job participants (such as contractor and subcontractor employees). The design professional is never wrong if he brings to the attention of the contractor and owner any unsafe construction practices he observes. No amount of responsibility for site safety on the part of the contractor will insulate from suit the design professional who is aware of unsafe conditions, yet fails to bring them to the attention of responsible parties before someone is injured. Indeed, under certain circumstances, the design professional with construction administration responsibilities may even be required to stop the work if risk to life or limb is imminent.

Number 3: Ignoring Professional Limitations

(I am an architect. I can do anything.)

All practicing design professionals recognize that the very act of taking up the pencil (or the mouse) to design is risky business. It is their confidence in their experience and competence that allows them to complete their projects. However, a healthy respect for one's own practical limitations is essential for professional growth while minimizing risk. This requires an architect to know when to "keep his ego in check," and seek help from others. If the design professional does not have significant expertise in a particular area, he should seek the help and advice of a person or firm experienced in that area. Enlisting the aid of a consultant spreads risk and is the act of a prudent person, indicating to any outside party (judge or jury, for instance) an intent to foster and control quality performance. There are other times when the architect should accept that an experienced contractor may, in fact, be more of an expert on a particular building type than he is.

This concern is greatest where the architect or engineer not only steps out of his realm of experience, but also out of the area of his specific competence or licensure (for instance, architects doing technical engineering calculations or engineers designing areas typically requiring building code knowledge). Even if it is not possible to defer the work or obtain the assistance of a consultant, the design professional should seek to document his efforts to learn the subject and apply the applicable building codes. Virtually every building system and material is supported by organizations with publications and online assistance. Some organizations offer interactive assistance and hotlines to answer questions about the proper use of the materials and building systems. Some municipal construction departments offer preliminary planning reviews to help the architect flush out sticky code compliance issues. Confirming the appropriate use of a product (especially one with which the architect is not experienced) not only increases the professional's fund of knowledge, but also helps to spread the risk if the product does not perform as anticipated. All of these tools are available to keep the design professional from straying out of his design comfort zone.

Number 2: Skipping the Subrogation Waiver Clause

(What's mine is yours)

Most modern boilerplate contracts (and front-end specifications) include a clause that requires parties suffering damage or casualty loss to submit the loss to their own insurance carrier for reimbursement rather than file suit to recover damages from any party to the project. This clause is more favorable to the architect and the contractor than the owner, as it is more likely that the work will be damaged than any damage will occur to the property of the contractor or his subs. To the extent that this provision carries over after substantial completion, it is even more certain that the owner's property, rather than the design professional's property, will be at risk. See the Risk Reduction Tools for a sample subrogation clause.

Therefore, it is essential that the architect make sure that the provision that limits the owner to sue only for those items for which he is not insured is included in the contract, as it can result in immunizing the architect from potentially millions of dollars of liability.

Number 1: Designing a Roof for a Multi-family Dwelling

(Like driving a defective car across thin ice)

Statistically, the building system most likely to fail, even in new construction, is the roof. Because they are high-performance components of the building, subject to the ravages of wind, sun, and rain, they must be well designed, installed, and maintained. When roof systems fail, interior damage results to finishes, drywall, ceiling systems, and sometimes equipment and furnishings as well. Damages resulting from roof system failure, therefore, are the greatest source of claims against architects' professional liability carriers, and the most likely cause of litigation against an architect.

If roofs are the most likely cause of a claim against an architect, the most likely parties to make the claims are condomium associations (also known as community associations). These groups are statistically more likely to sue an architect than any other. Therefore, it stands to reason that the design professional dramatically increases the likelihood of suit if he designs a roof for a multi-family building, whether or not it is intended that the building will eventually be owned by a condo association. Similarly, there is substantial risk in designing either a roof or a multi-family building independently.

The architect cannot even protect himself by limiting his multi-family residential design to developers or other less litigious clients than condominium associations. This is because residential complexes are bought and sold frequently. Buildings that start out as apartments often are the subject of condominium conversion efforts. This means that the individual units will be sold and, eventually, a condominium association will be formed. The first official act for the new association will be to hire a lawyer, an engineer, and a property manager, usually in that order. The attorney will often help the association to select an architect or engineer who will inspect the property. The original purpose of this inspection was harmless enough. The condo association's engineer was looking to determine that the property purchased by the unit owner (particularly the *common elements*—the part of the property, buildings and grounds, owned by everyone together) was consistent with the information in a public offering statement on file with the state or local governing authority. Unfortunately, this inspection has morphed into a full-blown review of code, design quality, and construction. In some cases, this property review has included the inspection of individual unit interiors. While this review may or may not expose the condo developer to financial risk, it certainly results in increased risk to everyone who ever designed, consulted, or constructed the buildings or any improvement to the property. The architect who does not have a statute of limitations tied to substantial completion is likely to find himself in a lawsuit instigated either by the condo association or the developer.

As mentioned, the most common problems are roof-related, since this is the building system most likely to suffer from the effects of the elements or poor maintenance. Examples of common professional liability claims include: poor structural design; insufficient insulation or ventilation; deficient construction; failure to flash, poor nailing technique, insufficient spacing of shingles; problematic products (e.g., fire retardant-treated plywood); and absence of ice and water shields to guard against ice damming. These are just a few of the problems associated with poor roof performance and leaks that are likely to turn up in a condo association claim against a design professional.

It is no wonder, then, that the system most likely to be the subject of a lawsuit is the roof. This will, hopefully, motivate the conscientious designer to make doubly certain of his roof design, particularly for code-mandated ventilation, ice, and water shields, as well as appropriate flashings at intersections, elevation changes, and other likely areas of water penetration.

It also requires the utmost precaution in the selection and retention of a good roofer with verifiable references. Depending on the size of the project, the architect may wish to make part of the contractor's contractual obligation that he pay for an independent roof inspector to ensure proper quality control with the installation, especially where the architect is unable or unwilling to regularly inspect roof conditions. The architect should select this inspector, and retain a copy of his report and maintenance recommendations for future reference.

There is, of course, one simpler and more effective means of avoiding condo association roof claims: do not design roofs for multifamily dwellings.

LIABILITY WARNING SIGNS

Sometimes people see problems a mile away and can take effective and timely action to avoid them. Other times, trouble strikes bolt-like, out of the blue, with little or no apparent warning. More often than not, the signs of trouble are present and palpable and the design professional is too busy or too much in denial to recognize their presence. Presented below are signs that the ship of professional liability avoidance is sailing on heavy seas.

Risk Reduction Tools

Sample Agreement Subrogation Clause

The architect and owner agree that each party shall carry appropriate property and casualty insurance coverage for their respective real, personal, or mixed property, whether such property is located at the jobsite or in other locations. The parties agree that that any damage sustained by either party, regardless of fault, shall be satisfied first from the proceeds of such insurance. No claim that is the subject of such insurance coverage will be the subject of any claim by the party suffering the loss or by its insurance carrier. Both parties further agree to require similar waivers of claims from their respective contractors, subcontractors, and/or subconsultants of whatever tier. This clause shall survive the project completion and be applicable as long as the parties maintain property or casualty coverage.

No Communication

Once the parties look for excuses not to communicate with each other, it takes a substantial effort to bring them back to a warm and happy place. Communication must be maintained, and the more disappointing or difficult the news, the greater the effort that must be made to effectively communicate. Even where people are uncomfortable at having face-to-face discussions, they must nevertheless keep the information flowing among them. This flow must include the relevant details of the project, including conditions, progress, questions, and decisions. When the communication flow breaks down, the project will suffer and liability will increase for the design professional.

When communication has broken down, it is critical to take immediate steps to restore communications before the impact to the project—and the professional—is irredeemable. An old architect's axiom says, "Call your client most when you least want to." Good advice, even it does come from an old architect.

Not Paying the Bills

Perhaps the surest sign of problems in a project is the failure to pay bills. This applies not merely to the owner's failures to pay the architect's invoices (although this one surely hits closest to home). It also applies to instances when the contractor fails to pay his subcontractors, suppliers, or others, or when the architect fails, without good cause, to pay his consultants. Each compensation lag is another leg up on the liability ladder. Firms that worry they may be stiffed for the work they have performed to date are naturally reluctant to perform future work in a timely and responsible manner. An unpaid subcontractor may not fully staff the next phase. An unpaid engineering consultant may be tardy in responding to RFIs or checking submittals.

Design professionals are often reluctant to discuss money issues with their client. This is unfortunate and potentially hazardous, because if the architect gets too far behind in payments, he loses perspective on the project and may eventually sink to the point where he literally gives up any hope of getting paid in full. The owner and architect start making bargains with each other (and with themselves), good money is thrown after bad, with the result that more money than was ever justified has been wasted on a project. The reality is that if the owner cannot finish the job, it is better to know sooner rather than later. Additionally, owners are sometimes not that much different than contractors in their attitude toward using other people's money. If a nefarious owner believes he can finish the job with someone else's money (in the form of credit that never gets repaid or liability suits as counterclaims for collection actions), he will do so. To avoid this type of owner, the architect should require the owner to pay his bills as they fall due. He must not make the problem worse by extending credit. As harsh as it sounds, architects should make sure to include a "no pay, no play" clause in their contracts to save themselves from a future as an indentured servant to the owner. Additionally, the design professional's desire to take the side of the owner in conflicts with the contractor should never extend into attempts to deprive the contractor of contractual monies fairly earned. Contractors who sense a lack of impartiality on the part of the architect may claim that the owner and the architect have entered into a conspiracy to maliciously defeat their honest attempt at getting paid. Improper and partial attempts to defeat the contractor's efforts to be paid (i.e., stalling payment application approvals, failing to approve applications on flimsy grounds) should not be considered. An architect who does engage in this type of conduct usually winds up with a new title—the defendant.

The design professional is rarely charged with an obligation to investigate what the contractor does with the funds he receives upon approval of the application for payment. However, where the architect learns of failures on the part of the contractor to pay his subs, suppliers, and vendors, this fact should be reported as soon as possible to the owner, as these are often signs of a contractor in distress and a project at risk.

Contractor Delays in Maintaining the Schedule

Few projects maintain perfect schedules. Weather, strikes, acts of God (and labor unions) can hold up the best prepared schedules. The design professional should be vigilant in watching for a contractor's consistent failure to meet project milestones. His first step should be to determine the reason for delay. Was it several days of bad weather or simply a failure of the contractor to have enough men or materials on site? Was the project delayed due to a recognized circumstance beyond the contractor's control, or the result of a failure to pay subcontractors on time or schedule follow-on trades closely enough? Any of these shortcomings are signs that the contractor needs reminders of his contractual obligations to get the project done within the time frames set in the contract documents. It is also a good time to reinforce with the contractor that the owner will not tolerate suggestions that holding the contractor to this obligation will be deemed an acceleration of the project schedule.

Adversarial Attitude

No one likes a bully. The natural tendency is to fight or flight, to push back, or to ignore. None of these tendencies are conducive to healthy project communication, or to healthy projects for that matter. An adversarial attitude from the contractor is a serious indication of trouble on the project. Smart, capable contractors want smooth and cooperative communication with the owner and architect. A contractor who is looking to create adversity on the job site is either foolish or foolish like a fox. The safest assumption an architect can make when he faces an adversarial contractor is to assume that the contractor is setting him and his client up for change orders or delay claims for later in the project. The presumption would be that the contractor hopes to bolster his case by harassing the architect into equally unprofessional behavior that would be viewed by an arbitrator or judge as hostile to the progress of the work.

In a more basic sense, adversarial relationships make it difficult to resolve the routine problems of construction, and make it more likely that otherwise solvable problems will either mutate into something more serious or be resolved in a manner less beneficial to the owner. Neither is acceptable, and the architect facing a continuously hostile contractor employee should ask the owner to intervene in demanding a change in personnel or attitude.

Slow Approvals on Change Order Requests

Owners do not like change orders. That is a basic fact, and they like them even less when they result from unclear scope or omissions in the construction documents. Some owners will delay signing off on recommended change orders for a variety of reasons:

1. The owner believes that delaying resolution will force the architect and contractor to settle the issue between themselves.

2. The owner considers the architect or contractor responsible for the change order, and rather than pay the change order and make a claim against either party, simply ignores the issue.

3. The contractor proceeds with work under dispute, and the owner thereby realizes that change orders can be delayed and settled later, perhaps at a discount to the claimed value.

Owners who would like to ignore change orders are rather common. Owners who successfully do so are only able to do so through the complicity of the architect and contractor. Architects must insist that owners face responsibility for change orders in a timely manner. If the owner feels the architect is liable for the change order, he has options through the architect's professional liability carrier to make a claim. The owner, however, is responsible for compensating the contractor for change order claims that his professional agrees are not reasonably a part of the contract for construction. Difficult and unpleasant though it may be, the architect must help the owner face this responsibility to the contractor.

Dangerous or Disorganized Work Site

Contractors are responsible for their work site. They know this, of course, and normally take charge of all things relating to its conditions. They realize that an organized and orderly job site is most conducive to schedule conformance and profit enhancement. An organized job site is a fundamental aspect of a well-run contracting operation. So when an architect sees a disorganized job site, it raises the concern that if the contractor is not successfully managing such a basic part of the project, is he effectively managing the more detailed aspects of the project? A poorly organized job site often is accompanied by substandard construction quality and haphazard schedule conformance. Worse yet is a disorganized project site that contains dangerous elements: poorly secured scaffolding, scattered debris, unmarked trenches or pits, construction debris and trash. Dangerous conditions on a job site are indicative of a contracting operation that is poorly managed and will struggle to complete the contract for construction on schedule and to the quality expected by the owner. Also, as discussed previously, the architect has no choice but to immediately raise safety concerns with the contractor and owner and press for correction of the unsafe conditions.

Late Submittals

Another sign of a poorly managed contracting business are late submittals. Late submittals mean the contractor is either shopping key components of the project well after signing the construction contract (a sign he bid the project with little or no profit), or that his office organization is understaffed or disorganized. Neither of these circumstances is good, because it means the contractor is imperiling the construction schedule. In the case of trying to recoup lost profit, the contractor is in double jeopardy since late submittals threaten the one route through which he might regain some money—advancing the construction schedule. Even if the architect cannot know the true cause of the late submittals, he should report to both the owner and contractor a concern that they imperil the success of the project.

Risk Reduction Tools

Best Ways to Stay Out of Trouble

1. Work under a good contract
2. Hire a good lawyer
3. Use a good professional liability insurance broker
4. Have access to good local contractor information
5. Maintain continuing education
6. Maintain a standard forms file
7. Write an introductory/explanatory letter
8. Practice continual calendar management
9. Utilize an information management system
10. Maintain membership in professional organizations

Excessive Requests for Information

Experienced architects can sense when requests for information (RFIs) are aimed less toward interpreting the documents than setting up a change order request. When numerous RFIs come in quickly, it is a clear signal that the construction documents are either poorly prepared or confusing to the contractor, or that he is scouring them looking for potential change orders. Both mean trouble for the architect, since even a well-intentioned contractor will feel compelled to submit change order requests for work that was not clearly indicated on the plans. A less well-intentioned contractor will use RFI responses as evidence to bolster change order requests for additional scope on even a well-prepared set of documents. This problem is found mostly in publicly-bid projects where the competition for work is more intense, and the pressure on contractors to bid projects with little profit is prevalent. It can also be found on private projects as well, particularly those where the contractor has mistakenly left scope out of his bid. It may also occur in cases where a particular subcontractor (usually one of the trade contractors) has had his proposal beaten down by the contractor or owner and is looking to recoup through change orders.

The Revolving Door

Contractors rely on their project manager (PM) and field superintendent to learn and manage their projects. It is very difficult for a new PM or field super to come into a project mid-stream, learn the details of the work, and carry it through to a successful completion. When a contractor changes personnel mid-stream, particularly a field superintendent, it could be a bad sign. High turnover among key construction staff is a sign of dissatisfaction and perhaps disorganization. Even when turnover occurs for innocuous reasons, the project may suffer as the new principals work to learn the project.

Staff changes may have a more troublesome meaning, however. Contractors struggling to build profit—or avoid losses—rely on these two key personnel to do so. If they are being replaced by others in the middle of the project for no compelling reason, it probably means the contractor is dissatisfied with their performance in managing the profitability of the project, or that his employees are unhappy with him. New personnel brought in will have clear instructions: perform the work for less money. Either way, they will have only a few options from which to choose in accomplishing this goal: perform the work faster and more efficiently, shop for less expensive subcontractors, recoup money through product substitutions, or claim change orders. From this bag of tricks, claiming change orders is the path of least resistance with the highest payback.

RISK REDUCTION TOOLKIT

Architects need to create their own toolkit to equip themselves with the means necessary to minimize the risk they create for themselves, and manage it effectively when they are thrust into problem projects. (See Figure 10-1 for the general keys to managing CA risk.) Here are the basic components of a risk management toolkit (see the Risk Reduction Tools).

1. ***A good contract:*** Most architects use only one or two standard contract forms. Work with an attorney to fine-tune a contract that works, and stick with it. The architect should incorporate by reference a standard set of terms and conditions.

2. *A good lawyer:* Owners and contractors have them, and so should design professionals. Any number of issues can prompt an architect to seek legal advice: contract language, threats of lawsuit or insurance claims, non-payment of invoices, and a host of other problems routinely faced by architects makes having access to a good attorney a necessity.

3. *A good professional liability broker:* Practicing without professional liability insurance is foolish and risky. Every design professional should obtain professional liability insurance and work with a broker who can provide contract review and other guidance to help manage risk.

4. *Access to good local contractor information:* Architects should talk to other architects, owners, and contractors to develop information on area contractors, their staffing, capabilities, and strengths. See the Risk Hazard Flags for contractor performance danger signals.

FIGURE 10-1
The cycle of effective construction administration.

Risk Reduction Tools

The Risk Reduction Toolkit

Reduce risk on a project by:

- Carrying professional liability insurance
- Using written agreements
- Defining architect services carefully
- Clearly stating excluded or additional services
- Setting budget and schedule early
- Obtaining owner signoffs for each phase
- Staying on top of receivables
- Warning owner of scope creep
- Creating quality construction documents
- Coordinating consultant documents with the architectural documents
- Utilizing in-house or outside expertise to check documents
- Using add and deduct alternates to create bid flexibility
- Carefully selecting bidding contractors
- Ensuring the contract for construction includes all bid addenda and negotiation agreements
- Responding to RFIs and submittals within the contract's required time
- Diligently reviewing applications for payment
- Preparing the punchlist with owner input

5. *Continuing education:* The practices of architecture and engineering change constantly. So also do forms of contracting and the technical aspects of specifying products. Continuing education is required for most state licenses and for AIA membership as well.

6. *Forms file:* Standard office forms, easily understood and used consistently by the staff, are the best way to ensure that project information is recorded in an organized manner throughout the course of the project—a fundamental component of risk man-

agement. The ready availability of professional practice forms and software makes it all the more likely that communication will occur and be recorded by the architect's personnel.

7. ***Introductory/explanatory letter:*** Begin the relationship with both owner and contractor with an introductory letter explaining the basic contractual obligations of each party and who will be responsible for meeting them.

8. ***Calendar management***

   a) Project and construction milestones: Maintain a project schedule for the owner, and review the contractor's construction schedule at each project meeting. Notify the owner of delays in the schedule. Urge the contractor to find ways to make up lost time.

   b) Payment requests: Applications for payment demand quick review and response. If additional information is needed, request it quickly. Relay to the owner within one week.

   c) Submittal and RFI responses: Architects should know the obligations of the owner/contractor agreement for submittal and RFI responses, and follow them.

9. ***Information management systems:*** Whether all paper, all electronic, or something in-between, the architect should

Risk Hazard Flags

Contract Performance Danger Signals

- Project behind schedule
- Workmanship quality issues
- Inadequate project staffing
- Late or chronically incomplete submittals
- Widespread substitution requests
- Invoicing ahead of completion
- Subcontractor or supplier payment complaints
- Frequent subcontractor changes

Risk Reduction Tools

Field Observation Key Information

- Date and time
- Weather and temperature
- Area of the work being observed
- Trades on site and work they are performing
- Specific problems noted/action taken
- Progress since last visit
- Safety conditions observed and reported to contractor

develop an office information management system that keeps project information of various types in an organized method that can be easily and quickly accessed. See the Risk Reduction Tools for recommended field observation report information.

10. ***National/local professional organizations:*** Every design professional should join his professional societies such as the American Institute of Architects and the Construction Specifications Institute. They provide valuable resources, continuing education opportunities, and the support and fellowship of other professionals in the field.

DOCUMENTATION

It is often said that a good professional keeps good records. These records include, at the very least, the communications received or issued by the professional. Communication methods have proliferated to the point where it is often difficult to keep track of what was said, by whom, when, and in what medium. No self-respecting architect hopeful of maintaining a communications link to the job can function without fax machines, wireless phones, and Internet access. He must also be prepared to preserve outgoing and incoming communications from all of these sources and under almost all circumstances. Because faxes come in writing and are delivered in "hard copy" form, they may be the easiest to maintain intact. Even e-mails are preservable with the click of the "print" button. While it may seem quaint, and even wasteful, to keep a paper copy of electronic communication, the need to preserve documentation on a project where an architect suspects trouble is brewing is so great that the redundancy of keeping records in two forms is a reasonable precaution. Project management software, some designed specifically for architects, is also available to provide integrated record-keeping for busy professional practices. See Figure 10-2 for recommended project documentation.

The most difficult thing to do is to consistently, accurately, and comprehensively preserve telephone discussions and face-to-face conversations. The architect may want to develop a desk-side form for recording essential information from even routine telephone calls, especially calls with owners and contractors. Affordable soft-

FIGURE 10-2
Critical project documentation.

ware is available to enable architects to easily create desktop electronic records of their phone conversations. However an architect records his verbal conversations, he should remember to identify who, what, where, when, and especially how any decisions were arrived at during the conversation. This information will pay huge dividends when trouble strikes. See the anecdote for one architect's story of the hazards of poor documentation.

Whether or not specifically tasked by either his contract or the front-end documents, the architect is often relied upon to be the "scribe" of the project. But the architect should make clear, especially in those instances in which the owner is using (and paying for) a contractor or construction manager to maintain the project records and documents, that while he, the architect, is not the repository of project documents, there are nevertheless documents that the architect must maintain as accurately and as comprehensively as possible.

The following specific types of communications call for additional considerations.

Requests for Information

Often a request for information is received from a party who might be seen as a potential adversary, such as a contractor. This is especially true for, but not necessarily limited to, projects where there is early acrimony. When writing to this potentially adverse party, or about them to a third party, regardless of the formality of the medium, do not forget that the correspondence may end up some day with a sticker on it reading "Exhibit A."

It goes without saying that responses to requests for information must be prompt or they will be featured in a delay claim. The design professional's response, however, must avoid telling the contractor *how* to build and should be restricted to design clarification, or *what* to build. Responses should be in complete sentences, either stated with the question or essentially restating it. Reference to the specific sheets, details, or specification sections, while appropriate and necessary, should not be the whole answer to the request for information. The design professional should fight the urge to tell the contractor how to build the project.

Submittals

When an architect receives a submittal from a contractor, it should be date-stamped to document when it reached his office. Contractor submittal logs show the date they sent the submittal and when they receive it back. Mail and transit times mean this log always shows a few more days available to the architect for review of the submittal than actually existed. For this reason, architects usually keep their own submittal log, with a record of when they actually received the submittal and the date they returned it. The architect's log documents that actual time the architect held the submittal, and tracks with normal contract requirements better than the contractor's log. When an architect returns a submittal to the contractor, he should do so using a transmittal letter. More than a form to document the date and contents of the return, the transmittal letter can be a valuable means for the architect to provide the general contractor with his concerns about the timeliness, scheduling, and overall professionalism of the submittal. Such a letter can form the basis of a defense to a claim by a contractor that the slow return of the submittal delayed the project. The shop drawings themselves, even effectively stamped and marked up, often do not convey to the people on a jury the difficulty experienced by the reviewer in dealing with late tendered, poorly conceived, and weakly executed submittals.

Site Observations and Project Meeting Minutes

The nature and content of site observations and reporting requirements are usually the result of the design professional's contractual scope of work and practice once construction has commenced. If the design professional visits the site regularly and is required to report on such visits, he should include the following information in each field report:

- Date, time, and duration of visit
- Weather conditions
- Specific portion of the facility toured
- Trades on site and the work they are performing

C.A. Anecdote
The Dimension Lawsuit
The Problem

Nick was served with the lawsuit on a Friday afternoon. By Monday morning he was sitting in his attorney's office, fidgeting while he wrote on a yellow legal pad everything he knew about the incident.

"So tell me what this is all about," said his attorney.

"About three months ago, in the middle of steel erection, my project architect visited the site and noticed that the mezzanine framing was about six inches short of where it needed to be. There was a dimensional error on our mezzanine plan, so if he had not caught the problem in the field, the mezzanine would not have met up with the wall construction below."

"It's good he noticed it when he did."

"Maybe not so good. My guy's idea was that the contractor would simply cantilever the steel deck beyond the framing an extra six inches and we'd figure out a way to run the wall framing up to it and created a finished look. He thought he had a verbal understanding with the field superintendent that he would take care of this. In his mind, at least, it was a simple, no-cost solution."

"What got us to this point?" asked the attorney.

"A few days later a revised shop sketch comes into the office showing additional miscellaneous steel framing to support the deck and create the cantilever. Our engineer agrees it's a good idea, and we approve it and return it to the contractor. My project architect noted under the approval stamp: "Approval does not authorize additional time or monies."

"Nice try. I take it the contractor wants both?"

"You got it. This little mezzanine extension has turned out to be an expensive proposition. The contractor sent us the change order and we rejected it, arguing that this error should have been caught by either his steel detailer or field superintendent, and that the problem could have been settled for no additional money by merely extending the steel deck as my guy suggested."

"But you approved the more extensive fix," said the attorney.

"Yes, but only if it didn't cost any extra money. We made that clear in the approval. He should not have performed the work if he intended to submit a change order for it."

"What does your client say?"

"He says someone made an error," replied Nick. "He doesn't care who pays for it, as long as it isn't him."

The Resolution

"You have a fundamental problem," said his attorney. "You have no documentation from the field by your project architect to support the contention that the contractor's field superintendent and he agreed to simply extend the deck. There is documentation to support that the contractor submitted a change in the scope and your office approved it. In our business, it only happened if it is recorded on paper. You are also trying to argue that you have the power to both require a change and determine its cost. You don't have that contractual power."

"But why did he make the change, knowing we were expecting it to be performed at no cost?" said Nick.

"Look, I'm not arguing there isn't some shared responsibility on the part of you and the contractor. Certainly both of you missed the blown dimension. He should have read your approval note and set off alarm bells before performing the work. You and your engineer should have pressed your argument that the decking could be extended at little or no cost. All of those things should have happened, and any of them could have avoided the lawsuit."

"Where does this leave me?"

"We better notify your professional liability insurance carrier. Since this is mostly remedial work, I think the owner will justifiably fight any additional costs on his part. The best we can do is work out a shared responsibility between you, the contractor, and his subs. We may want to drag your structural engineer's insurance carrier into this as well."

"All because of bad communication," said Nick.

- Specific problems observed or brought to the architect's attention, and actions taken

- Where appropriate, an update on the progress of the work since the last visit including all relevant building systems

- Observed safety-related issues, which should be brought promptly to the attention of the general contractor, in writing, with a copy to the owner. Immediate communication with the general contractor/subcontractors depends on the severity of and remedies for the problem.

If the design professional is the scribe for the project meeting minutes, he should consistently follow a recognized style and format for recording the events of the meeting, describing the nature of the issue, the action necessary to close out the item, the party principally charged with the action, and the follow up.

The communication of the minutes should be to the owner, the design team, and principal contractors. The minutes should always indicate the next scheduled meeting and set a reasonable deadline for reporting any alleged inaccuracies so the minutes can be finalized and everyone has been offered the chance to comment.

Proposals

Proposals to provide services are opportunities to show the design professional's unique ability to satisfy the client's needs at a reasonable price. Unfortunately, the proposal also exposes the professional to the snares of the "unintended contract." In these instances, the proposal, usually in the form of a signed letter, provides an outline scope of services and a cost for same. There is usually a signature line for the client indicating acceptance of the price and scope terms. This "offer-acceptance" format with essential (if not sufficient) conditions and the fresh signature of the offeree has all the earmarks of a contract which, in fact, it is. It is imperative that the architect's written proposal be labeled as such, along with a disclaimer that it is not intended to be a contract. The architect's proposal should clearly refer to a requirement that the parties will execute a contract to be negotiated by the parties, and note that the letter of agreement will be included as part of the contract. (For example: "This is a proposal only and not a contract. The parties expressly reserve the right to execute an agreement inclusive of all necessary terms and conditions, and including this proposal as an exhibit.")

Some design firms have shortened the process by incorporating by reference a standard list (one or two pages) of terms and conditions which, when added to the price and scope terms of the proposal, can serve in place of a formal agreement for professional services. Architects should take this course with caution and consult with an attorney to review the standard terms and conditions, if they desire to use this method.

SHARED RISK

The architect sometimes finds himself in the position of having committed an error or omission in his design, either out of exceeding his capabilities on a particular project or simply due to carelessness. In such an instance, the architect may have to bear responsibility for the error, including the costs of remediation and the consequential damages that may ultimately flow from the error. This can include damages due to delay to the contractor or the owner, especially if the owner can show that the delay cost him money for the substituted use of another facility or lost profits due to an inability to use the project to conduct his business pursuits.

However, it is just as often the case that the owner or contractor may wrongfully assert the negligence of the design professional and the project deficiencies may in fact have multiple causes. Listed below are some tips to alert the design professional to the possibility of errors on the part of other parties to the project and provide a method of analyzing the possibility of fault of project participants other than the architect.

Contractor

All parties must remember that it is the contractor and not the architect who is charged with the responsibility to construct the project. The architect has (hopefully) provided in the contract the limitations on his responsibilities in this area. The contractor is solely responsible for the means, methods, sequences, work procedures and techniques, safety procedures and programs, and for scheduling the work, not only his own but for his subcontractors and (often for) other prime contractors as well. The architect must first look to the specific nature of the owner's complaints and determine if the problem lies within the responsibilities of the contractor. The architect must ask such questions as:

- Is the problem the result of a failure of the contractor to follow the plans and specifications? Did the contractor make accurate and timely submittals of his work for the architect's review and approval?

- Is the item of construction a "performance specification" which calls for the contractor to design as well as build?

- Has the contractor inadvertently or intentionally substituted an out-of-specification product or system without first seeking permission for the substitution?

- Has the contractor completed work despite the express objection of the architect or failed to replace work that has already been rejected?

- Has the contractor utilized poor construction or management techniques in erecting the offending system?

These and a dozen other questions must be asked and answered to the satisfaction of the architect to get a clear understanding of the liability picture.

Construction Manager (CM)

Construction managers differ from the garden-variety general contractor in some fundamental ways. An agency CM (see the glossary for a definition) acts as an agent for the owner, and has a fundamentally different relationship with the architect and subcontractors as a result. The CM is not only the eyes and ears of the owner on the project, but often has very specific and technical responsibilities during both design and construction. As previously discussed, the CM often has constructability review responsibility, which entails several different, albeit related tasks. Such tasks include a thorough understanding of the owner's program, the capability of the architect's plans to satisfy the owner's legitimate expectations as expressed in the program, and the capacity of the plans to adequately articulate the design intent to the contractor in order to avoid change orders and unnecessary costs and delays. Here, the questions asked of the CM's performance, as compared to the contractor's, may be more subtle but they are no less important:

- Did the CM thoroughly understand the program mutually agreed upon by the owner and architect?

- Did the CM comment effectively and in a timely manner in cases where he believed the plans failed to articulate the program?

- Did the CM anticipate the likely areas of confusion, lack of clarity, or outright omissions that are being blamed for offending change orders?

- Prior to approving an error/omission change order, did the CM require the contractor or subcontractor to justify it as legitimate, rather than merely negotiate the price?

- Did the CM communicate with the architect in order to give him an opportunity to indicate where in the plans or specifications the appropriate comments placing the contractor on notice of his responsibility could be found? As an agent of the owner, the CM must be held accountable for his contractual obligations on the project.

Owner

The owner should expect to be held to the same level of scrutiny as the CM in those cases where he is managing his own project. Even where he has employed a CM, however, the owner still has specific responsibilities, and a deficiency in any of them may cause problems for his own project. To avoid problems relating on a construction management project, the architect must ask the following germane questions:

- Has the owner provided a complete picture of his needs and requirements for the project?

- Has he skimped on the details in order to save money?

- Has he been completely candid about his finances and his ability to fund a project of the type anticipated (including contingencies and emergencies)?

- Has he accurately anticipated his need for technology within the project or is this an afterthought?

- Has he listened to the admonishment of the architect about the folly of changing the design after construction has commenced or, for that matter, the project has been bid and contracts let?

- Has the owner provided for clear lines of communication, command, and control to minimize confusion?

- Has he stood by the architect to confirm his responsibility for design, notwithstanding the suggestion by the contractor to disregard the architect's advice?

- Has accurate backup information (legal, civil, geotechnical, accounting, and insurance) been provided, or has the absence of such information slowed the project?

- Has the owner made timely decisions regarding finishes and other items so materials could be ordered and long lead items received on time?

- Has the owner approved otherwise objectionable or out-of-specification work simply to advance the schedule?

- Has the owner failed to approve legitimate change orders for added scope or differing or changed site conditions which had the effect of delaying the work?

These and many other questions must be answered before any design professional willingly concedes responsibility for the damages claimed by the owner or contractor on any project.

Glossary

A

Addendum: Graphic or written additions to the bid documents, issued to the bidders of a construction project prior to the execution of the contract. Each addendum becomes part of the eventual contract for construction. The plural of addendum is addenda.

Add alternate: An alternate bid obtained by the owner for a limited addition to the scope of work. Add alternates are accepted or declined during the negotiation period with the contractor.

Additional services: Professional services provided by the architect or engineer, beyond the basic services required under his contract with the owner, provided for an additional fee.

Agreement: A written document stating the terms of the construction contract, including references to the documents that define the specific work to be performed under the contract as well as general conditions that define the right, responsibilities, and relationships of the various parties involved in the agreement. See also contract for construction.

Allowance: A specific value defined in the bid documents for contractors to include in their bid for a specifically-defined portion of the work. Allowances are usually used for defining material costs that have yet to be selected by the architect or owner. The final contract value is adjusted for the actual cost of the work covered by the allowance. Also known as a cash allowance.

Alternate bid: Amount stated in the bid documents to be added to, or deleted from, the base bid amount required under the documents if the alternate bid is accepted by the owner.

Application for payment: A contractor's written request for payment for work performed under the contract for a specific period of time. Applications for payment are usually accompanied by a schedule of values showing the percentage of work completed for the period in question. See also progress payment.

Approval: An architect's or an engineer's written acknowledgement that materials or equipment are acceptable for use in the work. Tacit approval may take the form of statements such as "No exceptions noted."

Approved equal: Where allowed under the construction documents, a product different from the one(s) specified by the architect or engineer that is deemed to be essentially equivalent to the one specified by the professional. Approved equal products carry the assumption that neither the owner nor the contractor is entitled to an adjustment in the value of the contract. Approved equal is a device used to prevent a competitive advantage for a proprietary specification.

As-built drawings: Construction drawings that are revised to show significant differences from the construction documents or the location of buried or hidden items. As-built drawings are maintained by the contractor throughout the course of construction and presented to the owner as part of the contractor's closeout documents. See also record drawings.

B

Back charge: A charge assessed by the architect or owner to recoup fees for professional time caused by another party's failure to perform his work in a professional manner (example: repeated reviews of improperly prepared submittals). Back charge also refers to the contractor assessing a charge against a subcontractor for extra costs the contractor incurred as a result of negligence or poor performance on the part of the sub.

Base bid: The total value bid by the contractor to perform the work required under the construction documents, not including work defined as a bid alternate under the documents. See also alternate bid.

Bid bond: A form of bid security, provided by the bidder, as evidence of a viable bid. If the contractor withdraws a low bid for a project, the owner may claim the bid security as compensation. Also known as bid security, bid guaranty, or a guaranty bond.

Bid conditions: A statement in a bid proposal that places limitations or conditions on the contractor's bid.

Bid documents: The complete set of documents distributed to bidders to use in preparing a bid for a construction project. These documents typically include: The invitation to bid, bid instructions, the construction drawings and specifications, addenda, and bid RFI responses.

Bid exclusions: Portions of the scope of work, either indicated or not in the construction documents, that a bidder specifically excludes from his proposal to the owner.

Bid form (also known as form of proposal): The document provided in the bid documents for the formal submission of the bid from the contractor.

Bid questions: Questions submitted during the bidding period by potential bidders seeking clarification from the architect regarding the construction documents. The costs associated with the responses to bid questions are assumed to be included in a bidder's proposal, and will ultimately be incorporated in the contract with the owner. Also known as bid RFIs.

Bidder: A person or entity submitting a proposal to construct a project in accordance with the bid documents. A bidder is not a contractor on a project until the owner executes a contract for construction with him.

Bidding and negotiation phase: A phase of architectural service in which the architect and his consultants provide support to the owner in analyzing the bid results and assisting the owner in negotiating with one or more bidders. Also known as procurement.

Bidding period: The length of time stated in the bid documents for the preparation of bids by the bidding contractors. The bidding period ends at the bid deadline.

Boilerplate: The front-end specification sections, consisting of the general and supplementary conditions, and conveying information specific to the contractor's operations on the site and relationships with other parties in the work.

Building code: The legal requirements established by governing agencies for minimum levels of construction.

Building code official: A representative of a code enforcement agency authorized to administer the enforcement of the building code in a particular area. Also known as a building inspector or construction code official.

Building permit: Approval issued by the appropriate local building or construction department allowing construction of the project to proceed. Building permits are often issued for various components of the work, including: demolition, foundation/footing, fire protection, electrical, and plumbing. Also known as a construction permit.

C

Certificate for payment: A statement from the architect to the owner confirming that the payment amount claimed by the contractor on his periodic application for payment is correct. The certificate is usually incorporated as part of the first page of the contractor's application for payment.

Certificate of occupancy: A document issued by the municipal building code official. The certificate of occupancy allows the building to be occupied for its intended use.

Certificate of substantial completion: A document issued by the architect on behalf of the owner stating the date that a building is suitable for occupancy by the owner for its intended purpose. This certificate is normally issued immediately following the granting of a certificate of occupancy by the local code official.

Change order: A written order from the owner and architect ordering a change in the scope of the work, and a corresponding change in the contract sum or contract time. A change order is an amendment to the contract for construction.

Change order proposal: A proposal from the contractor to the owner offering to perform additional work beyond that required by the contract for construction for a stipulated sum. One or more change order proposals are converted to change orders once they are signed by the owner and architect. Also known as a pending change order (PCO) or change order request (COR).

Changes in the work: Changes in the contract scope, time, and/or contract sum ordered by the owner. Changes in the work should be documented by change orders.

Closeout documents: Collection of documents required under the contract to be delivered prior to issuance of final payment by the owner. These documents typically include operation and maintenance manuals, as-built drawings, final release of liens, contractor's affidavit of payment of debts and claims, as well as warranty information.

Concealed work: Prior work performed by the contractor which is then concealed, or covered by subsequent work. Concealed work is a dispute issue when the owner or architect suspect the contractor is attempting to hide work that is not in conformance with the contract documents. Also known as covered work.

Construction administration phase: The final phase of architectural services, consisting of the architect's general administration of the contract for construction on behalf of the owner.

Construction budget: The sum of money stipulated by the owner as being available for the construction of the work—specifically the work represented in the bid documents.

Construction documents: The working drawings and the specifications.

Construction documents phase: A defined phase of architectural services in which the bulk of the construction documents (plans and specifications) are prepared.

Construction management (CM): A project delivery method in which a construction manager acts as an agent of the owner in procuring subcontractors and suppliers and performing the work for either a fixed or percentage fee based on the value of the work. CM work can be performed by parties who hold the contracts for construction (at risk CM), or parties who act strictly as agents for the owner (agency CM). A CM differs from a general contractor in his relationship with the owner and the subcontractors.

Consultant: Typically refers to an agent hired by the architect to provide specialized engineering, interior design, cost estimating, or other services as part of the overall professional services contract with the owner. May also refer to an agent hired by the owner to provide services directly to him.

Contingency (also known as contingency allowance): In construction management and guaranteed maximum price contracts, a percentage of the construction cost set aside for the contractor's use in completing the work required under the contract for construction. The contingency is normally managed and controlled by the construction manager.

Contract documents: The collection of documents, including drawings, specifications, bid instructions and form of proposals, bid addenda, request for information responses, and other written communication that are referenced in the construction agreement as obligations of the contractor in performing the work.

Contract for construction: The agreement between the owner and contracting entity (whether general contractor or construction manager at risk) for construction of the work represented in the construction documents. Also known as the construction contract, or the agreement between owner and contractor.

Contract time: The period of time, typically in calendar days, allowed for the contractor to complete the work described in the contract documents. Also known as the contract period, contract term, or construction time.

Contractor: See general contractor.

Contractor's affidavit of payment of debts and claims: A closeout document issued by the contractor at the end of the project certifying that he has paid all known claims from debtors for labor and materials related to the work.

Cost event log: A log normally maintained in construction management or guaranteed maximum price contracts in which the construction manager tracks known events that will increase or decrease the contingency or contract value.

Cost-plus agreement: A type of agreement between the owner and contractor in which the contractor charges the owner actual labor, material, and overhead costs, plus an agreed upon markup to cover overhead and/or profit. This type of agreement is often used in conjunction with a cap, or upset amount, to limit the owner's maximum cost. Also known as a cost-plus-fee agreement.

Credit: A type of change order in which the owner is credited with monies for work not necessary or removed from the contract for construction.

Cutting and patching: Miscellaneous work performed by the contractor or his subcontractors to run piping or cabling, to install portions of the work, to achieve proper fit and finish, or to install portions of the work in areas completed by another trade.

D

Date of commencement: The date on which the work required under the agreement is to commence. Absent any language to the contrary contained in the agreement, the date of commencement is assumed to be the date of agreement.

Date of substantial completion: The date on which the architect issues the certificate of substantial completion, or in the absence of a certificate, the date on which owner and contractor agree the work was substantially complete.

Deduct alternate: An alternate in the bid documents that provides for a selected portion of the work to be removed from the overall scope of the project, for the purpose of providing a means of reducing the cost of the project if necessary. The deduct alternate is accepted or rejected during the negotiation phase with contractor.

Design-build agreement: A single contract, or pair of phased contracts, where a single entity is responsible for both design and construction services, sometimes under a joint-venture basis.

Design development phase: The phase of architectural services following preliminary or schematic design, in which the architect describes the overall size and character of the project and sets the structure, systems, and superstructure of the design.

Design intent: In instances where the scope of the documents is in dispute between owner and contractor, the architect is called upon to offer an impartial statement of his design intent in creating the documents.

Due diligence: When applied to architectural practice, a legal principle requiring a person to exercise reasonable care, diligence, skill, and judgment in providing professional services. Also known as due care.

E

Errors and omissions (E&O): A type of professional liability coverage and claim associated with perceived mistakes or omissions in the construction documents.

Estimate of construction cost: A forecast of the construction cost at specified periods during the performance of the architect's services, usually prepared by the architect or a specialized construction cost estimate consultant.

Estimated quantity: A technique used in bid documents to require the contractor to include in his bid (and the contract for construction) a specific quantity of work that otherwise cannot be estimated (i.e., roof deck replacement).

Extra: A type of change order in which the contractor charges the owner additional sums for work that was not part of the contract for construction.

F

Fast-track construction: A method of project delivery in which construction work begins prior to final completion of the construction documents. Used on schedule-intensive projects, or in situations where material costs are rising rapidly.

Fee: The compensation for design professionals or construction managers on a project. While methods of pricing professional services vary, fee usually denotes a fixed amount for a stipulated level of services.

Field engineering: Specialized engineering provided by the contractor as part of the contract for construction, and related directly to the means and methods of construction, often for temporary facilities and systems. Examples of field engineering are: scaffolding, bracing, and shoring.

Field instruction: A written instruction from the architect to the contractor, instructing him to make changes to the work shown in the construction documents. A field instruction may or may not result in a change order or adjustment to the contract time. Also known as a field clarification or field order.

Field observation report: A report completed by a design professional during his periodic visits to a construction site to observe the progress of the work. Also known as a field report.

Field representative: An employee of the architect or one of his consultants, responsible for representing his firm in job meetings, job site observations of the work, and interactions with contractor personnel.

Field superintendent: The contractor's day-to-day site manager who is responsible for coordination of subcontractor activities on the site, job site safety, and expediting progress of the work.

Final acceptance: Following notification from the architect, acceptance by the owner that the project is complete in all regards and all closeout documents have been received. Final acceptance is normally a prerequisite to final payment.

Final completion: Notification from the contractor to the architect and owner that he has completed all obligations required under the contract for construction. The date of final completion determination is ultimately fixed by the architect.

Final inspection: Following final completion notification from the contractor, a general review of the site by the architect and owner to determine if any obvious reasons exist not to grant final acceptance of the work.

Final Payment: Following final acceptance, the last payment from the owner to the contractor, in which all retainage is returned and no contract sums remain to be paid.

Final release of liens: A part of the closeout documents on a project, in which the contractor releases the owner from any potential liens possible for the full value of the contract for construction, usually referencing contractors, subcontractors, and vendors.

Fixed limit of construction cost: As set by the owner, the maximum cost of the work that can be represented in the construction documents. Also known as the construction budget.

G

General conditions: The portion of the construction contract that describes the relationships, obligations, and rights of each party involved in the project, without regard to a specific trade or discipline.

General contractor: The prime or main contractor on a project, often responsible for coordinating and scheduling of other prime contractors. If there are multiple prime contracts on a project, the general contractor is the one responsible for general construction.

Guaranteed maximum price (GMP): The value established in an agreement between the owner and construction manager as the maximum price of the work represented in the contract documents, including labor, material, overhead, profit, and a contingency factor. Also known as guaranteed maximum cost or upset price.

H

Hidden conditions: Conditions on a project site, whose full extent is not known by the owner or design professional and cannot be fully defined in the construction documents. Sometimes confused with concealed work, which is new work covered by the contractor.

Hold harmless clause: See indemnification clause.

I

Indemnification clause: A contractual obligation requiring one party to protect another against damages or losses resulting from specific liabilities. Also known as a hold harmless clause.

Interpretation: A formal opinion issued by an architect when questions regarding the intent of the construction documents have been raised by the contractor. The architect is responsible for issuing an interpretation that fairly reflects his intent in creating the documents.

Initial decision maker (IDM): A term used in standard agreements created by the American Institute for Architects. Refers to an individual identified by the owner as having the authority to make day-to-day decisions relating to the progress of the work on behalf of the owner. This person may, or may not, have fiduciary responsibility. The initial decision maker may be the architect or engineer for the project, another professional engaged by the owner, or an employee of the owner.

Instructions to bidders: Instructions contained in the bid documents that tell the bidders how to prepare and submit their bids. A failure to follow the instructions, particularly in publicly-bid project, may invalidate a contractor's bid. Also known as a notice to bidders.

Insurance: Financial protection by an outside company for specific acts or perils, in specific amounts, required as part of the professional services agreement and/or the construction contract. Typically required insurance includes: automobile and general liability, builder's risk, employer's liability, professional liability, workmen's compensation, and loss of use.

Invitation to bid: The section of the bid documents inviting bidders to submit a proposal for the project. The invitation to bid typically includes bid deadline, pre-bid meeting, and other basic information relating to the bid submission.

K

Kick-off meeting: The first meeting held after the issuance of the notice to proceed, at which time the architect and contractor agree on means of communication, review the construction schedule and schedule of values, and set up a regular schedule of project meetings.

L

Letter of agreement: A legally binding letter stating the terms of an agreement between two parties. Often used in the form of a proposal from a design professional, signed as accepted by an owner.

Lien: There are two types of liens: mechanic's (labor) or material. A lien is a legal notice filed with a government entity that a debt for either the labor or material provided to a project has not been paid. A lien obligates that the debt be satisfied, or proper security provided, before the property can be sold or transferred.

Lien release: A document stating invoices from a particular subcontractor or supplier for labor or materials used on a project have been satisfied. Once a lien has been filed, its release can only be secured by the filing of a release or a court order.

Lien waiver: A document signed by a party that relinquishes his right to file a mechanic's lien or material lien against a project in exchange for payment for work performed. See also lien and lien release.

Liquidated damages: A clause in a construction contract stating that if the contractor does not complete the project within the contract period or other stated period, he will be subject to stipulated damages per business or calendar day. It represents the agreement of the parties to fix in advance the estimated amount of damages caused by the delay. If it is excessive or meant to intimidate the contractor into compliance, a court can invalidate it as a penalty.

M

Maintenance bond: A specific type of bond provided by the contractor to the owner guaranteeing to provide warranty work for a specified period.

Materials testing: Field testing performed by the contractor of specific elements of the work to demonstrate compliance with the specification requirements.

Means and methods: The particular equipment, operations, practices, procedures, and related activities used by a contractor to construct the work.

Mediation: An alternative method of voluntary dispute resolution, in which the parties to a dispute are guided by a trained mediator in arriving at an acceptable resolution to their differences. Often required as a necessary condition to litigation or binding arbitration.

Minor change in the work: A change of a minor nature, usually conveyed through a field order, that does not necessitate a change in contract value or contract time.

Mock-up: A temporary panel or model construction in the field to establish the expected quality, color, and installation levels acceptable for the project. Also known as a representative sample.

Modification to the contract documents: A written change in the construction documents issued by the architect in the form of a field instruction, change order, or response to a request for information.

N

Non-conforming work: Work performed by the contractor that does not meet the intent of the construction documents. The contractor may be required by the owner to remove and replace non-conforming work at no additional cost to the owner.

Notice of award: Written notice to a bidder that his proposal has been accepted and the owner intends to enter into a formal contract for construction with him.

Notice to bidders: A document included in the bid documents for a project, notifying bidders of the opportunity to bid on a project and summarizing the procedures for doing so.

Notice to proceed: Written notice from the owner to the contractor authorizing him to proceed with the construction and establishing the beginning point of the contract time.

O

Observation of the work: The function performed by the architect during the construction period by periodically visiting the site and observing the general progress and quality of the construction in relation to the construction documents.

Open-shop: A project that is constructed by non-union labor. Also known as merit shop.

Operation and maintenance manuals (O&M manuals): Closeout documents submitted by the contractor that convey the information provided by the manufacturer for equipment and other installations in the project. These documents may also include subcontractor information regarding the details of custom or modified installations.

Out-of-scope work: Work acknowledged by the owner to be outside the requirements of the contract for construction.

Outline specifications: Preliminary specifications prepared by the architect, usually during the schematic design or design development phases of a project, establishing a minimum performance level for building systems, superstructure, and finishes.

Owner: The firm or person financing the project and contracting with design professionals and contractors as his agents in meeting his facility needs. The owner may or may not hold title to the location of the construction.

Owner representative: A person employed by the owner to represent his interests during construction of the project. This person may or may not have fiduciary responsibility to the owner or the ability to make decisions directly for the owner. Also known as an owner's inspector, initial decision-maker, project manager, or clerk-of-the-works.

Owner's own forces: Agents hired by the owner to perform specialized portions of the work outside the contract for construction. The contractor is often required to coordinate with the owner's forces, but is not otherwise contractually bound to them.

P

Partial release of liens: A lien release accompanying a periodic application for payment that releases the owner from liens only up to the current point of payments in the project. The owner is not protected from liens for work performed beyond the current application. Also known as a periodic release of lien. Also see final release of liens.

Payment bond: A bond that requires a surety to insure payment of the labor and material costs required by a contractor to perform the work required in the contract for construction. A payment bond is often combined with a performance bond.

Penalty clause: Incorrect, out-of-date, and misleading term (no longer use in construction contracts) for liquidated damages assessed against a contractor for exceeding the contract time specified in his agreement. See liquidated damages.

Percentage fee: Compensation for a construction manager or architect based on an agreed-upon percentage of the construction cost.

Performance bond: A bond in which the surety guarantees to the owner that, should the contractor default, all work required under the contract will be performed by the surety. The performance bond is usually combined with a material and labor payment bond.

Performance specifications: Direction provided in the project manual (specifications) instructing a contractor to provide a specific building system or product (including the design, fabrication, and installation), based on a set of performance criteria identified in the project manual. The contractor is often required to submit the design for the system in the form of a drawings, signed and sealed by a licensed professional, to the architect or his consultants for review. Examples of performance specification systems include: temporary facilities and devices, shoring and bracing, pre-engineered metal buildings, roof systems, and certain mechanical and electrical components of the building.

Phasing: The organization of the construction work into separate stages, each one preceding the other, usually dictated by site, occupancy, or monetary restrictions.

Pre-bid conference: A conference conducted by the owner or the architect with potential bidders to discuss bid requirements and instructions, and collect bidders questions regarding the bid documents.

Preliminary design: Drawings prepared during the early stages of a design project. Preliminary design follows schematic design and precedes design development in the order of architectural services to a client.

Prime contract: One of the construction contracts with the owner, in which the contractor is responsible a portion of the work. The term is often used in publicly-bid projects to encourage competition and lower contract bids. See also general contractor.

Program: A written list of areas, relationships, and criteria developed by the architect and owner to describe the requirements of the project.

Progress payment: A periodic application for payment submitted by the contractor for work performed to a certain date. Also known as a periodic application for payment.

Progress schedule: An interim schedule produced by the contractor at regular project meetings, showing the progress to a certain date in relation to the construction schedule approved by the owner.

Project budget: The sum of money stipulated by the owner as being available for the entire project, including the construction budget, legal and financial costs, land acquisition costs, furnishings, and professional services costs. See also construction budget.

Project cost: Total final cost of the project, including land acquisition, legal and financial costs, professional services, and construction.

Project manager: Typically, the employee of the contractor or construction manager responsible for the overall management of the contractor's operations on the site and fulfillment of contractual obligations with the owner. The architect's employee responsible for the professional services of the project may also be referred to as a project manager, as may an owner's representative responsible for representing the owner's interests on the site.

Project manual: The specifications prepared by the architect. A component of the construction documents, bid documents, and contract documents.

Project meeting: A meeting held on the site with representatives of the owner, contractor, and architect to review construction progress and issues affecting completion of the contract for construction.

Project representative: A term used to refer to either the owner's agent in representing his interests on the site and serving as an initial decision maker (IDM) for the owner, or the architect's employee responsible for providing construction administration services to the owner.

Proposal: An offer to perform specified design services or construction for a specified price. See also bid form, change order proposal, and request for proposal.

Punchlist: A list prepared by the architect or owner's representative near the end of the project, listing those portions of the work not yet completed or not performed in accordance with the construction documents. The punchlist is often a pre-condition of substantial and final completion.

Q

Quality reference standards: A section of the specifications listing construction standards for various components of the work as published by national organizations. Ideally, specific references are made to each standard in the applicable specification section.

Quote: Typically, a confidential bid from a subcontractor to a contractor during the bidding period, but also a proposal during the construction period for change order work. Also known as a quotation.

R

Record drawings: An updated set of construction documents, prepared by the architect from as-built information supplied by the contractor.

Rejection of the work: The act of a design professional in informing the contractor and owner that he considers specific portions of the work to be non-compliant with the contract for construction. Rejection of the work is often the basis of disputes during construction.

Request for information (RFI): A request to the architect or his consultants from the contractor seeking additional information or clarification of information contained in the construction documents, or seeking the architect's instructions for what to do in response to field conditions that differ from those anticipated in the construction documents. The contractor usually tracks RFI submissions on an RFI log.

Request for proposal (RFP): A written request from a project owner that a design professional submit a response which includes the scope of work and fee for a defined project. Often, the design professional is invited to provide a statement of qualifications (including the resumes of key personnel), and examples of prior projects.

Retainage: A percentage of the value of the work performed to date withheld from each progress payment to ensure contractually required performance, and returned to the contractor near or after the completion of the project. Also known as retention or retained percentage.

S

Sample: A physical example of a product (typically a finish product or color sample) submitted by the contractor for review by the architect or owner. See also mock-up.

Schedule: A document prepared by the contractor prior to start of the work showing the detailed steps and milestone dates necessary to complete the contract for construction within the contract time.

Schedule of values: A statement provided by the contractor that allots the contract value for specific portions of the work, usually following the Construction Specifications Institute (CSI) format. A schedule of values usually accompanies each application for payment, showing the percentage of work completed for the period in question.

Schematic design phase: After programming, the phase of architectural services in which the architect prepares general design studies, often in the form of elevations and floor plans, to assess the most desirable means of achieving the owner's goals for the project.

Scope of work: All the building systems comprising the work to be constructed by the contractor. The entirety of the project represented in the contract for construction, or other more limited documents such as a change orders or field instructions.

Shop drawings: Detailed drawings for various components of the work (i.e. structural steel, trusses, ductwork), prepared by subcontractors and submitted to the architect and his consultants for by the contractor review and comment. Shop drawings and other submittals are submitted according to a schedule to facilitate the orderly and efficient flow of the work. See also submittals.

Site: The location referenced in the contract for construction where the work takes place. Also known as the field, the job site, the project site, or the work site.

Special conditions: A section of the specifications prepared by the architect setting forth unique or unusual conditions specific to the project, and affecting the contractor's work. Examples: Limited working times, site access, noise or dust limitations.

Specifications: The written portion of the construction documents, which describes the specific products and quality standards required for the construction of the work. See also performance specifications.

Standards of professional practice: Statements of ethical practices created by professional societies (such as the American Institute of Architects or the American Society of Civil Engineers) to guide their members in the conduct of their professional practices. Also known as standards of professional care.

Statement of probable construction cost: Cost estimates prepared by a design professional at various stages of the design of a project to provide guidance to the owner in the anticipated costs of the work. See also estimate of construction cost.

Statute of limitations: A statute stating a time limit, commencing with the date of injury or damage, in which legal action must be brought, otherwise the right may be lost. Statutes of limitations vary by state and type of legal action.

Statute of repose: A statute stating the maximum dates to bring claims for building projects, commencing from the date of occupancy or substantial completion. Statutes of repose differ from statutes of limitation in several regards: statutes of repose begin from a certain date, allow fewer exceptions, and usually allow a longer time period for the filing of claims.

Stipulated sum agreement: A construction contract that stipulates the owner will pay the contractor a fixed sum of money for completion of the work represented in the contract documents. Typically, the sum can change due to a change in scope, additional work, or a change in circumstances. Also known as a lump-sum agreement.

Stop work order: An instruction issued by the owner to stop work on a project. Typically issued in extreme situations only, to address safety, gross construction deficiencies, weather, monetary, or labor situations. Also, an order issued by a regulatory government agency under similar circumstances.

Subcontractor: A secondary contractor engaged by a general or prime contractor to perform a selected part of the work required by the contract for construction.

Submittal: Documentation demonstrating compliance with the construction documents submitted by the contractor to the architect for review and approval. See also shop drawings.

Subrogation clause: The substitution of one firm or person for another in a legal contract. Construction agreements typically restrict this activity. Also, the right of an insurer who pays an indemnity to seek compensation from the entity covering the loss.

Substantial completion: The date certified by the architect when the building, or a portion of the building, is sufficiently complete to be used or occupied by the owner for its intended purpose. This date may, but does not necessarily, coincide with the issuance of a certificate of occupancy by the local building code official, particularly when an owner's special requirements supersede the requirements of the building code. See also certificate of substantial completion.

Substitution: A product or process proposed by the contractor in lieu of the specified product or process, offered either with a credit for a lesser product or with no credit for an equivalent product. Typically, a substitution must be expressly approved by the architect prior to implementation or installation.

Supplementary conditions: A section of the specifications prepared by the architect that modifies the general or special conditions of construction.

Supplier: An individual or firm supplying construction products to the work, but not providing labor for the installation of those products. Also known as a material supplier, material man, or vendor.

Surety: A company authorized by law to guarantee that the work required under the contract will, depending on the terms of the bond, be performed in whole or in part by securing payment for labor and materials. Sureties are also commonly referred to as bonding companies.

Surety bond: A legally enforceable instrument of guarantee where one party (the surety) agrees to be responsible to another party (the owner) for the debts, obligations, or contractual responsibilities of a third party (the contractor—also referred to as the principal in bonds).

Survey: In land surveying terms: a drawing prepared by a licensed land surveyor to determine the physical boundaries, characteristics, and limitations of a site, including (dependent on the type of survey ordered): property boundaries, utility and easement locations, topographic features, vegetation and existing trees, area of the property, and adjacent properties and roadways.

In architectural terms, a building survey represents the act of documenting the existing conditions of a building, including overall dimensions, interior layout, elevation and fenestration information, and basic structure and systems documentation.

T

Time and materials agreement: A form of construction pricing where the contractor is compensated for the actual labor and materials costs incurred, plus an agreed upon markup for overhead and profit. Also known as T&M.

Time is of the essence clause: A provision in a contract alerting a responding party that a delay in resolving an issue or responding to a request, particularly where contractual time limitations apply, may result in a change order or schedule delay claim or other forms of damages.

Trade union: An organization that represents groups of skilled laborers. The union organizes its workers and bargains on their behalf for a labor agreement (collective bargaining agreement) that fixes the wages, benefits, and working conditions of its members. Construction contracts can specify the use of union labor, or the payment of union wages and benefits non-union workers. Also known as a union or labor union.

Turnkey agreement: A type of project delivery method similar to design-build agreements, with the exception being that in turnkey agreements the contractor may provide land, financing, design, and construction, retaining ownership of the project until a completed facility is transferred to the owner. A variant of turnkey agreements is sometimes used in public projects to avoid the strictures of public bidding laws.

U

Unit prices: Amounts required in the bid proposal or stated in the contract for construction as fixed for a specific measure of product (i.e., square foot of masonry). Unit prices are used to provide the owner with a competitive price for adding additional scope to the project. Sometimes used in conjunction with estimated quantities.

V

Value-engineering: The process of substituting less expensive materials and systems to reduce costs in a project without reducing scope.

W

Waiver: A document that voluntarily abandons a legal claim or right, usually in exchange for a service or product.

Warranty period: The period of time after completion of the contract during which the contractor is responsible for repairs at his own expense to any equipment or installations that do not perform properly, but without regard to fault. This period is typically set at one year from the date of final acceptance.

Work: The entire scope of the project as defined in the contract for construction.

Working drawings: The drawings documenting the work to be performed, which, when combined with the specifications, form the construction documents.

Z

Zoning permit: A permit issued by a governmental agency in an area permitting the placement, though not the construction, of a building on a site for the purpose or use intended. A zoning permit is usually a condition of receiving a construction or building permit.

Index

Made in United States
North Haven, CT
21 February 2023

32944500R00176